Hazardous Waste Minimization Handbook

Thomas E. Higgins

LEWIS PUBLISHERS

Library of Congress Cataloging-in-Publication Data

Higgins, Thomas E., 1948–
  Hazardous waste minimization handbook.

  Bibliography: p.
  Includes index.
  1. Waste minimization—Handbooks, manuals, etc. I. Title.
TD793.9.H54 1989   628.5′1   88-38019
ISBN 0-87371-176-9

LEWIS PUBLISHERS, INC.
121 South Main Street, Chelsea, Michigan 48118

PRINTED IN THE UNITED STATES OF AMERICA   5 6 7 8 9 0

To Judy and Nate

About the Author

Thomas Higgins is a senior environmental engineer in CH2M HILL's Industrial Water and Wastewater Treatment Department. He received a BS in Civil Engineering and an MS and PhD in Environmental Engineering from the University of Notre Dame. Before joining CH2M HILL, Dr. Higgins was an Associate Professor of Civil Engineering at Arizona State University, where he taught courses in environmental engineering. He also consulted and performed research on physical and chemical wastewater treatment processes. Dr. Higgins is senior technical adviser for CH2M HILL projects that involve metal plating and finishing processes and waste treatment, hazardous waste minimization, or the physicochemical treatment of industrial or hazardous wastes. He is currently senior advisor for waste treatment plant evaluation and design projects for Martin Marietta Aerospace, Raytheon, Bell Communications Research, M/A-COM, NASA Marshall Space Flight Center, and General Dynamics. Dr. Higgins was project manager for a study of industrial process and waste treatment innovations that have resulted in the reduction of hazardous waste generation at Department of Defense installations and is program manager for a task order contract with the Environmental Policy Directorate of the Department of Defense. He is a registered professional engineer in eight states, has received numerous national awards, and has written more than 45 papers on the subject of environmental engineering.

About CH2M HILL

In the mid-1930s, three Oregon State College engineering students—Holly Cornell, Jim Howland, and Burke Hayes—talked with one of their professors, Fred Merryfield, about starting an engineering firm some day. Following graduation, these students went on to graduate school and then to military service, where they served as engineers during World War II. Though separated, the four continued to correspond and plan. In January 1946 in Corvallis, Oregon, the partnership of Cornell, Howland, Hayes, and Merryfield was formed, specializing in providing sanitary engineering design services for Oregon's Willamette Valley.

In the late 1940s, the name CH2M came about. Some say a client suggested it; others remember it resulting from a word game. Development of innovative treatment techniques, such as multimedia filtration, led to the design of the world's first advanced wastewater treatment plant at Lake Tahoe. The Tahoe project led to a merger with Clair A. Hill and Associates of Redding, California, in 1971, and the addition of HILL to the firm's name.

Throughout its history the firm has enjoyed steady growth geographically, expanding to more than 45 offices with 3,400 employees working in more than 50 disciplines that provide services to water and wastewater, solid and hazardous waste, industrial, transportation, and energy clients. An important part of many industrial projects has been providing waste minimization services, concentrating or reducing waste disposal costs, and recycling material. CH2M HILL was recently ranked as the third largest architectural/engineering firm in the country, and the largest doing primarily environmental engineering work.

Since 1946, CH2M HILL has provided professional services in management, engineering, architecture, planning, economics, and environmental sciences throughout the United States and overseas. The company is entirely employee-owned.

Foreword

Past hazardous waste disposal practices have led to such dramatic environmental stories as Love Canal and Times Beach—once thriving communities, now ghost towns. Congress responded to Love Canal with the Superfund program, designed to clean up waste or at least reduce the health and environmental effects from uncontrolled hazardous waste disposal practices and prevent new uncontrolled hazardous waste sites from developing.

Both Superfund and RCRA regulate by mandating waste management and disposal practices for wastes that are already produced. Congress and the public have come to realize that such programs, while helpful, do nothing to reduce the amount of wastes still being produced.

A new approach was needed and the Congress and state legislatures began to mandate the minimizing of hazardous waste produced during manufacturing rather than relying on containment for protection. In this context, waste minimization has come to mean the modification of manufacturing processes to reduce waste generation.

There is a strong economic as well as regulatory argument for minimizing wastes. Disposal costs are escalating rapidly because of increasingly stringent and restrictive treatment and disposal requirements. A further and compelling reason for industry to adopt minimum waste-producing techniques is to avoid the potential liability that attaches to hazardous wastes once produced.

Currently, Congress is considering many additional approaches to waste minimization. While legislation and regulations can provide requirements for waste minimization, they cannot assist a manufacturer in setting up a minimization program or selecting changes that will work in a particular plant. Dr. Higgins' new *Hazardous Waste Minimization Handbook* provides that assistance.

This book is organized around the industrial processes common to most manufacturers that produce the bulk of hazardous wastes at individual facilities: solvent cleaning and metal preparation processes; machining and metalworking; metal plating and surface finishing; painting; paint stripping; and industrial waste treatment. It is filled with practical ideas, examples, and case studies—both successes and failures—based on both Dr. Higgins' personal experience and projects conducted by CH2M HILL. Case studies include process flow diagrams and cost information for capital as well as for operating and maintenance costs.

In summary, the *Hazardous Waste Minimization Handbook* provides practical hands-on guidance for corporate and plant managers and their environmental staffs for implementing a successful waste minimization program. It is also a ready source of ideas and techniques for reducing waste in specific manufacturing processes. For many, the book should become their primary reference on waste minimization.

William D. Ruckelshaus
Former Administrator, U. S. Environmental Protection Agency

Preface

This book, designed to assist industrial engineers and managers in making changes in purchasing, manufacturing, and waste handling practices to reduce the costs and liabilities of waste disposal, begins by defining waste minimization in the first two chapters. A description of the economic and regulatory incentives a company has for setting up a waste minimization program follows, and experiences of the author and CH2M HILL with industrial clients and their successful and unsuccessful projects are related. Based on these experiences, a method is outlined for companies to use to establish a waste minimization program and implement individual projects.

Chapters 3 through 7 of this book describe specific waste reduction techniques applied by a number of industries. The chapters in this section have been organized around industrial processes (i.e., painting, metal finishing, machining, etc.) rather than around standard industrial classifications (i.e., automotive, chemicals, etc.). Industrial processes selected for inclusion are those that are commonly employed in many of the industrial sectors and that contribute the greatest volumes of hazardous waste at a typical facility: machining, cleaning, plating and surface finishing, paint and coating application, and paint and coating removal. These processes and the types of waste produced by each are discussed. Methods used to reduce waste generation are described and illustrated with case studies. Sources of equipment, design data, and installation and operating costs are provided.

The book concludes with treatment methods that can be employed to reduce the volume or toxicity of waste, thus further reducing the costs of disposal.

The organization chosen for the book follows the methods that should be employed by a company when implementing a waste minimization program. First, an organization needs to be established to coordinate waste minimization activities, to demonstrate that the corporation is serious about the program, and to select projects for funding and management support. Next, the book describes methods by which industrial processes can be modified or waste recycled to reduce or eliminate wastes needing treatment or disposal. The final section deals with waste treatment to further reduce the volume or toxicity of wastes so that minimal waste has to be disposed of with minimal liability and risk to human health and the environment.

Acknowledgments

The genesis for this book was a CH2M HILL report for the Department of Defense evaluating waste minimization efforts at government-owned industrial facilities. The findings and case studies in that report have been reorganized and significantly expanded by the inclusion of materials from numerous other CH2M HILL waste minimization projects. The author wishes to acknowledge the following individuals and their contributions to the preparation of materials for this book.

Name	Chapters	Contribution(s)
Linda Tymciw		Editing
Kevin Hayes		Graphics
Drew Desher	5	Metal Plating, Surface Finishing
Stephen Graham	3	Machining and Other Metalworking Operations
Tom Card	4,6	Solvent Cleaning, Painting
Ben Fergus	4	Solvent Recycle
Rick Johnson	8	Thermal Destruction
Randy Peterson	7	Paint Stripping
Bill Cobb	1	Economic Analysis

The author also gratefully acknowledges the many unnamed CH2M HILL staff members who have assisted in the preparation of this book or worked on the projects used as examples.

Contents

List of Figures

Figure

List of Tables

Table

List of Case Studies

Case Study

CHAPTER 1

Introduction

WHY SET UP A WASTE MINIMIZATION PROGRAM?

The Hazardous and Solid Waste Amendments (HSWA) of 1984 have made it more difficult and expensive to dispose of hazardous wastes. Some wastes have been or will be banned from land disposal entirely (e.g., liquid wastes), and other disposal practices are severely restricted. Restrictions have caused a shortage of approved facilities, yielding higher cost of disposal. Also, past generators are today's potential responsible parties (PRPs) with legal liability even if following today's regulations. Also, generators of hazardous waste are required to certify that they have instituted a waste minimization program. In addition, financial and legal incentives make reducing or entirely eliminating the generation of hazardous waste increasingly attractive.

REGULATORY IMPERATIVES FOR WASTE MINIMIZATION

Land disposal has been the standard means used by industries to dispose of solid and hazardous wastes. Improper or inadequate containment of these wastes has resulted in contamination of air, water, and land resources. In response to a number of notable incidents of contamination (e.g., Love Canal), Congress has mandated that waste production be minimized, and that whatever cannot be eliminated, be disposed of in a safe manner.

Waste minimization has been a long-term Congressional goal, as is evident in most environmental legislation. The Federal Water Pollution Control Act of 1972 set a goal of "zero discharge" of wastes to the nation's waters by 1985. The original hazardous waste act was entitled the "Resource Conservation and Recovery Act" (RCRA), not the "Hazardous Waste Disposal Act of 1976." In the Hazardous and Solid Waste Amendments (HSWA) to RCRA in 1984, Congress stated its policy:

> The Congress hereby declares that it is to be a national policy of the United States that, where feasible, the generation of hazardous waste is to be reduced or eliminated as expeditiously as possible. Waste nonetheless generated should be treated, stored,

1

or disposed of so as to minimize the present and future threat to human health and the environment.[1]

As part of HSWA, generators of hazardous waste were required as of September 1, 1985, to certify on manifests that:

A program is in place to reduce the volume or quantity and toxicity of hazardous waste determined to be economically practicable.

and that:

The proposed method of treatment, storage, or disposal will minimize the present and future threat to human health and the environment.

However, neither HSWA nor subsequent regulations define what constitutes a waste minimization program in either the law or subsequent regulations. It requires only that such a program be established and that waste minimization be evaluated. In the future, documentation of such programs will be required.

Complicating the disposal of hazardous waste are the so-called "land bans" portions of HSWA in which, in subsequent years, land disposal of certain wastes is either restricted or completely prohibited. HSWA has banned the disposal of "free liquids" in landfills and requires that certain wastes (solvents) be incinerated. These regulatory decisions have resulted in requirements for waste minimization because complying with the regulations requires either pretreatment (solidification to eliminate free liquids) or disposal methods (incineration of solvents) that are more expensive than land disposal.

WHAT IS WASTE MINIMIZATION?

Trying to answer this question has stirred up considerable controversy. As initially defined by the U.S. Environmental Protection Agency (EPA), waste minimization includes anything that reduces the load on hazardous waste treatment, storage, or disposal facilities by reducing the quantity or the toxicity of hazardous wastes.

Several committees of Congress requested that their research arm, the U.S. Congressional Office of Technology Assessment (OTA), determine how successful industries had been in reducing the production of hazardous waste. This study is important to industry because the results are being used by Congress to draft additional legislation to increase the pace of waste minimization.

The resulting OTA report[2] took the position that "real waste reduction" was accomplished only through in-plant changes. The evaluation included only those changes that reduced the generation of waste in the production process itself. It did not include efforts to reduce the volume or toxicity of wastes once they had been generated. The report found that even though there were considerable economic incentives for waste reduction, little effective waste minimization had been accomplished (based on its narrow definition). The authors also recognized that "it would be extraordinarily difficult for government to set and enforce waste reduction standards for a myriad of industrial processes." The report, therefore, recommended that a program that emphasized incentives and technology transfer be favored over a regulatory approach to waste reduction.

The EPA would be responsible for enforcement of any regulatory legislation on waste minimization. In a report requested by Congress,[3] EPA concluded that additional regulatory legislation would not be required before 1990 but that industrial efforts should be monitored between now and then to determine if a regulatory approach was needed. Within its existing organization, EPA has implemented additional programs to encourage state minimization programs and to provide assistance on technology development and transfer.

As the chief regulatory agency responsible for protection of human health and the environment, the EPA has established methods that it recommends to industries to reduce the unfavorable impacts of hazardous waste disposal. EPA's priorities have been ranked as follows:

1. waste reduction

2. separation and concentration

3. exchange or recovery

4. incineration or other treatment

5. security of ultimate disposal

In neither the EPA report nor the OTA report is there currently a recommendation to mandate minimization standards. However, it is likely that additional legislation for waste minimization will be passed by Congress before 1990.

ECONOMIC INCENTIVES

In industry, there is a general perception that pollution control is a cost of doing business and that money or effort expended in compliance should be minimized. A marginally profitable company may be tempted to say that it cannot afford to pay for effective waste disposal, let alone develop an effective waste minimization program.

It has been said that "Wastes are resources that are out of place." Anderson[4] developed an economic model of production, blending the traditional theory of productive services with the concept of material processing energy use and waste generation. He provides a schematic of the production process in Figure 1.1. Two points are noteworthy: (1) there are three outputs, including Q (the amount of produced goods), W_E (the amount of energy consumption wastes) and W_M (the waste from the incomplete use of material inputs); and (2) the amount of material input equals the production output ($R = Q + W_M$).

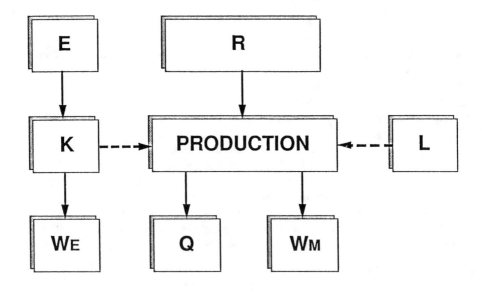

Where: K = Fund of Capital
          E = Stock of Energy Inputs
          Q = Stock of Output Product
          Wм = Stock of Wastes from the Materials Inputs
          L = Fund of Labor
          R = Stock of Material Inputs
          Wе = Stock of Wastes from the Energy Inputs

Figure 1.1.   An economic model of the production process (Source: Reference 4).

A firm's cost function can be determined using the inputs and outputs in Figure 1.1.

$$TC = P_1R + P_2E + P_3W_M + P_4W_E$$

Where: TC = total cost; P_1 is the price of input material; P_2 is the energy price; P_3 and P_4 are waste disposal costs; and R, E, W_M, and W_E are the amounts of a particular output.

If a firm wishes to minimize its total cost, it should be intuitively obvious to the reader that both disposal costs and waste amounts are a good place to start cutting costs. Disposal costs can include legal and permitting costs relating to short-term disposal practices. If disposal costs escalate (and they have been), then reducing the amount of generated waste should be investigated. There can also be long-term liabilities (i.e., Superfund) associated with waste disposal practices. It generally is difficult to reduce input amounts (assuming there is direct negative effect on revenues) or input costs (assuming competitive pricing is available for these input goods).

Using rigorous microeconomic techniques, a firm's production function can be modeled when $W_M = 0$ and when $W_M > 0$. From a microeconomics perspective, there is an optimal level of W_M that should be produced for a given production process; Anderson provides two graphical representations of this. This optimal level is based on trade-offs within the production process: inputs versus outputs. Waste reduction can be achieved by varying the production process, nature of the product, or nature of the input material.

A number of companies have found that, rather than increasing costs, waste minimization programs have actually reduced operating costs.

Example. The 3M Corporation has had a ''Pollution Prevention Pays'' program since 1975.[5] In 10 years, the 1,500 projects supported under this program resulted in a savings to the company of more than $235 million ''. . . for pollution control facilities that did not have to be constructed, for reduced pollution control operating costs, and for retained sales of products. . . .''

Example. A study performed by CH2M HILL for the Department of Defense[6] found that conversion to a dry media paint stripping technique at Hill Air Force Base resulted in an annual savings of $5 million. If the method were adopted by all three branches of the military, DOD has estimated that the annual savings would exceed $100 million.[7]

Similar results have been achieved by both large and small industries.[8-10] The common finding is that waste minimization can (and should) be justified on a cost-reduction basis.

ECONOMIC ANALYSIS OF WASTE MINIMIZATION

While many potential waste minimization projects can result in an economic benefit to the firm, this is not always the case. It is important to note that the Congress requires waste minimization to an ''economically practicable'' level, not to total elimination. As described by Anderson, there can be an economically optimum level of waste material generated by a given production process.

Cost-benefit analysis is the traditional method of selecting between environmental alternatives; in this case, between waste generation (or minimization) alternatives. It is typically used more by government and academia than the private sector (the private sector generally uses a modified version).

The first step of cost-benefit analysis is relatively straightforward. The costs of the waste minimization project are determined, including capital, labor, materials, and external (such as waste disposal). Capital costs can be calculated on an initial or annualized basis.

The second step is somewhat harder, in that the benefits of a waste minimization program must be estimated. Benefits include the value of any marketed good that might be produced, the value of nonmarketed goods (such as annualized savings of future liabilities or current disposal costs), and the value of any secondary benefits (i.e., public relations, community relations, or some other "public" benefit). These benefits are then summed up. Determining the time benefit of nonmarketed goods and secondary benefits is extremely difficult and is typically pursued by academicians rather than private sector analysts.

Costs are then compared to benefits on either an annualized or present worth basis. The waste minimization alternative with the highest net benefit should be implemented. However, if there are limited benefits, there is another economic evaluation method that can be used. In comparing alternatives that have only costs and possibly some salvage value, the rate of return for a given individual alternative is usually negative. For a rate of return analysis of alternatives that provide a service (such as waste minimization), you must analyze the *incremental differences* between the alternatives to determine if the incremental investment in more costly alternatives is justified by incremental savings. The following example illustrates this type of analysis.

Example. A company currently has Process A that has annual operating costs (OC) of $7.5 million and waste disposal costs (DC) of $1.5 million. This company is considering switching to Process B, which has an initial capital cost (C) of $14 million, annual operating costs (OC) of $6.5 million, and waste disposal costs (DC) of $1 million. Disposal costs are anticipated to rise 10% per year. The company would like a 15% rate of return over a 10-year life, assuming no salvage value.

Negative incremental operating costs are the same as positive savings, as shown in Table 1.1.

Another method of comparing alternatives A and B is by Net Present Value:

At i = 15%
NPV_A = $-$58,493
NPV_B = $-$53,796
$DNPV_B - NPV_A = +$4697$

Or, Process B is the least costly alternative over the 10-year life.

A simpler cost analysis method, payback period, is often sufficient for simple decisions. For the example given, the difference in annual costs between the existing (A) and the proposed alternative (B) is $1.5 million. The capital expenditure of $14 million is paid back in 9.3 years. This payback is often sufficient to justify small projects where the probable payback period is short enough (less than 5 years) to reduce the impacts of ignoring inflation or the time value of money.

An effective waste minimization program can be established based on the goal of increasing profits. Profits rise when a company reduces the costs of production or of waste disposal, or when it increases the value of production. Regulations have increased and will probably continue to increase the cost of disposing of materials with the most potential to harm human health and the environment when deposited

Table 1.1.   Present Worth Equations (in Thousands of Dollars)

Solution:

		OC = 6500	OC = 6500	OC = 6500	OC = 6500
Process B	C = 14,000	DC = 1000	DC = 1100	DC = 1210	DC = 2360
	0	1	2	3 - - - - - - 10 yrs.	

	OC = 7500	OC = 7500	OC = 7500	OC = 7500	OC = 7500
Process A	DC = 1500	DC = 1650	DC = 1815	DC = 2000	DC = 3890
	0	1	2	3 - - - - - - 10 yrs.	

		ΔOC = 1000	ΔOC = 1000	ΔOC = 1000	ΔOC = 1000
Process B–A	ΔC = 5000	ΔDC = 650	ΔDC = 715	ΔDC = 790	ΔDC = 1530
	0	1	2	3 - - - - - - 10 yrs.	

$$\$5000 = \$1650\ (P/F_{i/1}) + \$1715\ (P/F_{i/2}) + \$1790\ (P/F_{i/3}) + \$2530\ (P/F_{i/10})$$

By trial and error:
   for i = 20%   Present Worth = \$7980 – \$5000 = + \$2980
   for i = 30%   Present Worth = \$5739 – \$5000 = + \$739
   for i = 40%   Present Worth = \$4389 – \$5000 = – \$611
   by interpolation i = (739/1350) × 10% + 30% = 35.5%

in landfills. Therefore, basing waste minimization programs on cost-effectiveness (by including immediate and long-term waste disposal costs) is of value both in itself and because it meets the intent of Congress to reduce the volume and toxicity of wastes being disposed of in landfills.

METHODS USED TO MINIMIZE WASTE

Several methods have been successful in accomplishing waste minimization. They usually fall into one of the following categories:

- Change materials purchasing and control methods.
- Improve housekeeping practices.
- Change production methods.
- Substitute less toxic materials.
- Reduce wastewater flows.
- Segregate wastes.
- Recycle or reclaim wastes.

- Treat waste to reduce volume and/or toxicity.

- Delist wastes that are not toxic.

A short description of each of these methods follows, as well as illustrations of how these changes have resulted in waste and cost reduction.

Change Materials Purchasing and Control Methods

Frequently, raw materials and supplies are bought based on minimizing purchase cost without considering disposal cost.

To minimize disposal of new or unused material:

- Reduce to a minimum the number of different products (i.e., cleaning fluids, cutting oils, etc.) used. This streamlining mitigates shelf-life problems and reduces the number of partially used containers to dispose of.

- Buy in container sizes appropriate to the actual use. It can be less expensive to buy quart containers of a perishable product than to purchase gallons or drums of that product at a lower unit cost and later have to dispose of the unused portion.

- Reduce the inventory of hazardous materials to a minimum, and ensure that old containers are rotated from the back of shelves to the front when new material is purchased. This effort will reduce the disposal of new materials whose shelf-life has expired.

These measures will reduce the cost of raw materials and waste disposal, as well as the investment tied up in working capital inventory.

Example. At one large repair and rework facility, the majority of hazardous waste collected was either partially used material or unopened containers of new material that had exceeded its shelf-life. At the same facility, a solvent-recycle program was rendered useless by a purchasing agent who continued to deliver the same quantity of new solvent to the users on a scheduled basis, regardless of the need for new material. Excess purchasing resulted in increased disposal of older virgin solvent to make room for subsequent shipments.[6]

Improve Housekeeping Practices

Excessive waste production often results from sloppy housekeeping practices. Leaking tanks, valves, or pumps can cause process chemicals to spill onto the floor, requiring cleanup and disposal of wastes. Poor cleaning of parts before placing them in process tanks can greatly reduce the useful lives of process chemicals and can increase both the volume of waste requiring disposal and the cost of chemical replacement.

Example. At two similar plating shops, hard-to-replace parts were hard-chrome plated in preparation for machining and reassembly. At one shop, the plating baths were dumped on an average of once a year at a cost of $25,000 per year. At the other shop, only one plating bath had been dumped in 27 years of operation. The principal reason given for the difference was careful precleaning and masking of parts prior to plating.[6]

Key advantages of housekeeping changes are that they can usually be implemented quickly, they require little if any capital investment, and they are likely to result in substantial reduction in waste production.

Change Production Methods

There is a tendency to continue using the same manufacturing process even after improved methods have been developed. The old saying, "If it ain't broke, don't fix it," comes to mind. However, sometimes the process can be broken in other than a production-related sense if it produces too much waste. Frequently, changes that result in waste reduction can result in production improvements as well.

Example. For more than 50 years, the Navy has been using the same methods for hard-chrome plating. In the meantime, improved methods have been developed in the Cleveland area. By creating a program to reduce the waste from chrome plating operations, the Navy updated the entire chrome plating process and incorporated the improved methods developed in Cleveland. The result was a zero-discharge plating process that also produced a more uniform plate while using a considerably smaller tank area.[6]

Substitute Less Toxic Materials

Frequently, a nontoxic material can be substituted for one that is toxic or that causes a special waste treatment problem. The reduced cost of disposal or the reduced exposure of workers to a toxic material can justify the change.

Example. Cadmium was traditionally plated from a bath containing cyanide. Cyanide wastes require separate treatment from other plating wastes, thereby increasing the complexity and cost of waste treatment. When Lockheed successfully switched from an alkaline cyanide bath to an acidic noncyanide cadmium bath, plating quality was equivalent and overall costs were reduced. Reduced costs resulted from simplified treatment, despite a higher cost for plating chemicals.[6]

Reduce Wastewater Flows

Reducing the volume of wastewater can reduce disposal costs. If the volume is reduced sufficiently (and concentration increased proportionally), it is feasible to recover useful materials from the waste.

Example. At the Pensacola Naval Air Rework Facility's plating shop, a recirculating spray rinse system was installed on the hard-chrome plating line. Rinse volumes were reduced to less than the evaporation rate from the plating bath. The rinsewater was returned to the plating bath, aiding in recovery of plating chemicals and resulting in the elimination of rinsewater discharges.[6]

Segregate Wastes

Wastes that may require unusual treatment or disposal should be separated from other wastes. The rule for hazardous waste mixture states that mixing a regulated hazardous waste with a nonhazardous waste renders the whole mixture legally hazardous. Also, recycling a material usually requires that the waste remain as clean as possible prior to reuse.

Example. For solvent recycling using simple stills to be feasible, individual solvents must be segregated. At one facility, an attempt to recover heptane from a used storage tank containing waste heptane was abandoned when it was found that the reclaimed heptane failed to meet specifications. This failure was caused by mixing of waste solvents in this tank.[6]

Recycle/Reclaim/Reuse

It is often less expensive to recycle a chemical than it is to purchase new material and pay for disposal costs. Sometimes a material may no longer meet the specifications for a process in which it is being used; however, the material may still be suitable for other uses at the facility.

Example. High-purity freons are used for cleaning liquefied gas lines at the Marshall Space Flight Center. The Center will collect and use these solvents for general cleaning at the facility, rather than continuing the present practice of offsite disposal. A plant producing a waste acid stream purchased caustic to neutralize the acid for disposal with the plant's wastewater. The same plant produced alkaline sludges that were disposed of in a hazardous waste landfill. By neutralizing the waste alkaline sludges with the waste acid, they were able to dispose of the sludges in a commercial (nonhazardous waste) landfill. The result was reduced raw material costs and reduced disposal costs.

Treat to Reduce Volume and/or Toxicity

Disposal costs are based on the classification of wastes and on their volume. If a waste has hazardous characteristics, then disposal in a hazardous waste disposal facility is required. If the waste is treated to eliminate these hazardous characteristics, it can be disposed of in an industrial waste landfill, which is considerably less expensive. Transportation and disposal costs can be reduced by reducing the volume.

One means of treating waste to reduce its volume and toxicity is incineration. Incineration can accomplish the following tasks:

- reduce volume of waste to inert ash.

- change the character of waste by combusting organics.

- change the state of waste by evaporating and combusting liquids, leaving only solid residues.

- provide treatment of gases and vapors.

By burning the organic or combustible fraction of wastes, the volume of waste for ultimate disposal can be reduced to the minimum inorganic content or ash. A volume reduction of 70% to 95% is readily achievable for industrial or office rubbish. This combustion can reduce the apparent toxicity of the waste caused by organics or can change the character of the waste. Evaporating liquids and burning the combustible fraction of the liquid will leave behind only the inorganic ash for disposal.

An incinerator can also provide direct treatment of gases or vapors in air. A contaminated air stream can be ducted directly to a burner block, mixed with fuel, and burned, thus destroying the organics. This technique avoids water treatment for the discharge from a wet scrubber or solids disposal from carbon adsorbers. Besides reducing the volume or toxicity of wastes, incineration can provide a means for recovering energy or minerals. By using energy recovery techniques, a combustion system can provide steam and/or electricity.

Mineral acids can be recovered from the combustion of waste materials. Burning chlorinated solvents generates hydrochloric acid (HCl), which can be recovered in an absorption tower as an 18% to 20% solution in water. Wastes with a high concentration of sulfur can be burned and the sulfur recovered as sulfuric acid. This latter process requires catalyst beds and absorbers in series.

Other minerals can be recovered by burning solid waste streams. The gold from scrap circuit boards can be recovered by burning the resins. By burning scrap sandpaper, sandpaper grit can be recovered and recycled.

"Delist" Wastes That Are Not Toxic

A waste can be "listed," or regulated, as hazardous because it comes from a process that has generally been found to produce hazardous wastes. This regulation may force a company to dispose of a waste in a hazardous waste disposal facility even though the waste does not actually have hazardous characteristics. There is a regulatory procedure by which a listed waste can be delisted if it can be demonstrated that the waste does not exhibit hazardous characteristics. Since the delisting procedure can be expensive and time-consuming, it should be used only when the cost savings would be significant.

REFERENCES

1. "Resource Conservation and Recovery Act," Section 1003(b), as amended by the Hazardous and Solid Waste Amendments, Public Law 98-616 (November 1984).
2. "Serious Reduction of Hazardous Waste," U.S. Office of Technology Assessment, Washington, D.C., OTA-ITE-317 (September 1986).
3. "Report to Congress: Minimization of Hazardous Waste," U.S. Environmental Protection Agency, Washington, D.C. (October 1986).
4. Anderson, C. L. "The Production Process: Inputs & Wastes," *J. Environ. Econ. Management*, 14(1) (1987).
5. Susag, R. H. "Waste Minimization: The Economic Imperative," *Waste Minimization Manual*. Government Institutes, Inc., Washington, D.C. (1987), pp. 2–7.
6. Higgins, T. E. "Industrial Processes to Reduce the Generation of Hazardous Waste at DOD Facilities," Phase 2 Report, NTIS No. AD-A159-239 (July 1985).
7. The Assistant Secretary of Defense for Manpower, Installations, and Logistics, "The 1984 Annual Environmental Protection Summary" (1984).
8. Huisingh, D., L. Martin, H. Hilger, and N. Seldman. "Proven Profits from Pollution Prevention: Case Studies in Resource Conservation and Waste Reduction," Institute for Local Self-Reliance, Washington, D.C. (1986).
9. Campbell, M. E., and W. M. Glenn. "Profit from Pollution Prevention: A Guide to Industrial Waste Reduction and Recycling," Pollution Probe Foundation, Toronto, Ontario (1982).
10. Overcash, M. R. *Techniques for Industrial Pollution Prevention* (Chelsea, MI: Lewis Publishers, 1986).

CHAPTER 2

Setting Up a Program for Waste Minimization

LEARNING FROM SUCCESSES AND FAILURES

In the last few years, waste minimization has been the theme of several publications and conferences. At the conferences, the most interesting and useful papers are often the case studies that document experiences in applying particular technologies. However, almost without exception, the presentations describe only success stories. Thus, not only do these reports tend to give a false impression that success can be achieved no matter what modification is tried, they also, by not describing unsuccessful projects, prevent the reader from learning from the mistakes of others.

If a company has the same technical problem and similar production facilities as those cited in this book, the success stories will provide useful information. However, success stories may not show why a particular waste minimization program succeeded or provide the information necessary to develop a successful waste minimization program. Therefore, since frequently one can learn enough from examining failures to know what not to try and to look for variations that might prove successful, examples of failures are also included in this book.

EXPERIENCE FROM WASTE MINIMIZATION PROJECTS

At CH2M HILL, the author managed a study for the Department of Defense (DOD), in which the effectiveness of 42 industrial waste minimization projects was evaluated. These projects had been proposed and many of them implemented at government-owned industrial facilities. Facilities included those that are government-owned and company-operated (GOCOs) and those that are government-owned and government-operated (GOGOs).

The purposes of the CH2M HILL study were to:

- Identify the reasons why some of the projects succeeded while other, apparently similar, projects failed. The study identified not just technical factors but also institutional and motivational factors that contributed to the outcome.

- Evaluate the wider applicability of the three most successful projects at similar DOD facilities.

- Provide technology transfer within and among the services by providing hands-on workshops for three of the most successful and widely applicable projects.

- Improve the long-term success of DOD's waste minimization programs by identifying the mechanisms by which successful projects can be identified and implemented.

Some of these projects involved great ideas that failed to be implemented because of insurmountable institutional barriers or because they lacked an individual sufficiently motivated to spearhead the change. Some projects were implemented on the wrong process or in a situation in which the volume of waste generated was insignificant compared to the effort required to make the change. However, in several projects, the benefits from reduced waste generation exceeded the costs of implementation. In addition, three projects were so successful that their replication throughout DOD could significantly reduce both waste generation and overall operating costs.

Since completing that study, the author has had the opportunity to work with other CH2M HILL clients in developing waste minimization strategies for their industrial facilities. From these experiences and from discussions with successful waste minimization coordinators, the author has come to some conclusions regarding organizational and technical methods that proved successful, as well as those that did not.

ESSENTIAL ELEMENTS OF SUCCESSFUL MINIMIZATION PROJECTS

Based on a background of evaluating waste minimization projects, the following conclusions are offered. Although there were specific circumstances and reasons behind the success or failure of each waste minimization project, successful projects had two characteristics in common:

- Production personnel were strongly motivated to implement and maintain the necessary changes.

- Technologies used were "elegant in their simplicity."

These two characteristics were integral parts of each successful project; moreover, at least one of these elements was missing in each failure. The two characteristics are expanded into a number of factors in Table 2.1. Some factors are elements of both motivation and technology (i.e., some technical improvements provide motivation for the change). Duplicate factors are listed and discussed only once. Some discussions include examples. These examples and the respective technologies are described in more detail in the later chapters of this book.

Top Management Effectively Supports the Project

For successful minimization programs, support is provided at a sufficiently high management level to influence production and environmental policy decisions. Fre-

Table 2.1.   Motivational and Technical Factors for Successful Waste Minimization

Motivational Factors

Top management effectively supports the project.
Rewards and recognition are provided for successful minimization efforts.
Operations personnel are involved in planning.
A "Champion" implements the project.
Implementing the program reduces production costs.
Implemention of the program increases productivity.
Waste disposal costs are charged to production units.
Production people appreciate environmental impacts of waste disposal.

Technical Factors

Implementation of the program improves product quality.
Change is "evolutionary" rather than "revolutionary."
Equipment is adapted to local conditions.
New equipment is simple to operate, reliable, and easy to maintain.
Extensive training is provided for operating personnel.

quently, waste disposal and environmental protection are viewed as service functions that are subservient to production. Successful modification usually requires use of production resources to provide environmental protection. In a cost-cutting and staff-reduction environment, it is usually difficult to justify adding personnel for environmental protection. The following example shows how one individual coped with this personnel problem.

Example. At Robins Air Force Base in a large aircraft repair facility, an individual set up a profitable 10-person solvent recycling program. Because he assumed the function of supplying, handling, and disposing of solvents for individual shops, management in exchange allowed him a partial staffing slot from each shop served.

Rewards and Recognition Are Provided
for Successful Minimization Efforts

An incentive program for submitting ideas is useful only if the "good ideas" for potential waste minimization projects are successfully implemented and maintained. In addition to rewarding those who originate the ideas, an effective program also provides rewards and recognition for individuals who do the hard work of implementing projects and making them succeed. Well-publicized rewards and recognition provide an incentive to other workers.

Operating Personnel Were Involved in Planning

In successful programs, many operating personnel are involved in the design and installation of the system to obtain their input and to inspire them to adopt the process change. Since production people operate and maintain the modified process, they should have a say in designing a system that will meet their needs.

A "Champion" Implements the Project

Many successful projects are led by a Champion who strongly believes in the modification, ramrods the project, and overcomes development and startup problems. A Champion has to overcome the inertia that protects an existing process that "works" (although it may produce an excessive quantity of waste). Even the most difficult-to-operate process can be made to work by an individual who is too stubborn to believe that it cannot be done.

Example. This highly successful solvent recycling program was set up by a former Kentucky "moonshiner" who would not accept the "truism" that central recycling programs do not work. He was able to convince management to buy stills and assign him manpower slots to recycle solvents. He then built a solar still to recover contaminated jet fuel for use in ground equipment.

Implementation of the Program Reduces Production Costs

Production managers are usually evaluated on the basis of their control of production costs, product yields, and product quality. Modifications that result in reduced costs for the individual production unit are therefore more likely to gain local support than projects that result in reduction of waste disposal costs, which are usually shared company-wide—if at all—through reduction in overhead charges.

Implementation of the Program Increases Productivity

Reducing personnel requirements, thus increasing productivity, is a strong incentive for change.

Example. The plastic media blasting (PMB) process for stripping paint was originally developed by a technician at Hill Air Force Base. The technician wanted to eliminate the use of methylene chloride and phenolic paint strippers to improve working conditions for his coworkers. In developing the process, the Air Force discovered that the new process reduced personnel requirements for paint stripping an aircraft from 240 work-hours over a week to 24 work-hours over a 6-hour period. That productivity improvement (and cost savings) is now the driving force behind the rapid development of the process in the Air Force.

Waste Disposal Costs Are Charged to Production Units

At facilities where modifications were successful, production people were aware of the true costs of hazardous waste disposal and considered those costs when making decisions to implement waste minimization programs.

Example. At one facility, more than $100,000 was invested in an elaborate vapor recompression evaporator for recovering chrome from the rinsewater used in the limited production of hard-chrome plated parts. Waste production numbers for the facility were not available to justify the modification, so the facility used numbers from a high-volume, decorative-chrome plating shop as an estimate. After much difficulty in getting the process to operate, the rinsewater was analyzed, revealing that only $100 worth of chrome per year was being lost in the waste stream in the first place. Costs for operating the minimization system were greater than for operating a conventional waste treatment facility for chrome removal.

Production People Appreciate Environmental Impacts of Waste Disposal

Most people are motivated by a desire not to pollute the areas in which they work and their families live. Letting workers know the effects of what they throw away can motivate them to be more careful.

Example. One of CH2M HILL's chemical engineers spent the early part of her career as an environmental coordinator for a major aircraft manufacturer in the Pacific Northwest. Solvents kept turning up in the storm drains. She traced the source to several floor drains in the shop area, into which workers were pouring used solvent. She had signs made that read "NEXT STOP IS YOUR FAUCET AT HOME." The violations stopped. When asked by workers what they could do to reduce waste, her response was, "Act as though your mother were looking over your shoulder."

Implementation of the Program Improves Product Quality

More and more frequently, the quality of American products is an important basis for selection by the end user or consumer. When production changes to reduce waste production also improve the quality of the products, they tend to be successfully implemented. Conversely, changes that decrease the quality are soon abandoned.

Example. A zero-discharge chrome plating process was successfully implemented at a number of Navy shops, principally because the overall package of improvements included improved uniformity of the plate. As a result, subsequent machining and rejection of parts was considerably reduced.

Change Is "Evolutionary" Rather Than "Revolutionary"

On successful projects, off-the-shelf equipment is adapted to a new application; special or complex equipment is avoided. The greater the number of modifications attempted at the same time and the more experimental or less developed the equipment is, the greater the likelihood that the process will prove unreliable and be abandoned in favor of an existing method that has "all of the bugs worked out of it."

Example. PMB paint stripping was successful partly because the modification used sand-blasting equipment, a technology that is well developed. The plastic media used for the new method causes less wear and tear on the equipment than sand; hence, little modification was required. The modifications that were adopted were to make the equipment lighter and easier to use, taking advantage of the media's reduced aggressiveness.

Equipment Is Adapted to Local Conditions

Care is taken to tailor the modification (even for off-the-shelf equipment) before transferring it to facilities where it has not been tested. Soliciting (and using) suggestions from the local operators helps to ensure that they will ''buy into'' the change. If a suggestion is not used, care is taken to explain that decision to the individual who made the suggestion.

Example. A zero discharge chrome plating system has been successfully adapted at a number of Navy plating shops, based on the general principle of reduced rinsewater use and return of concentrated rinsewater to the plating tank to make up for evaporation. At one facility, the platers adopted use of recirculating spray rinse to reduce rinsewater use. At another facility the platers chose to use countercurrent multiple rinses. Both processes accomplished the goal of reducing rinsewater use and recovery of metal to the plating tank, but each used a method adapted to the capabilities and desires of the individual users.

New Equipment Is Simple to Operate, Reliable, and Easy to Maintain

Successful modifications are straightforward and simple to operate, thus requiring minimal training of personnel. If a new process is unreliable and reduces production rates, the old methods are quickly reinstated. Maintenance requirements for the new equipment are minimal for successful programs.

Example. An unsuccessful system for recovery of chrome from plating rinsewater employed both ion exchange and vapor recompression evaporation. The resulting equipment was complex, difficult to operate, and frequently shut down for maintenance. This system was soon replaced by the simpler, more successful zero-discharge plating system of the previous example.

Extensive Training Is Provided for Operating Personnel

Minimization projects usually require operational personnel to learn how to operate new equipment or how to operate existing equipment by new methods. Successful projects include training for more operators than are required, thereby ensuring that backups are available when needed. Sometimes unsuccessful projects are abandoned when key personnel leave the facility without training a replacement.

Example. Initially, only a few platers were trained to use the new zero-discharge chrome plating system, and management counted on them to train others. Unfortunately, the trained platers moved on to other jobs before they were able to train others. The project was salvaged by setting up a new training program in which all platers are trained in the new operating methods. At a successful plating shop at Cherry Point Naval Aircraft Depot, time is provided for the best plater to run a training program for the less experienced platers.

TIPS FOR SETTING UP A SUCCESSFUL WASTE MINIMIZATION PROGRAM

Based on experience with successful and unsuccessful waste minimization projects, the following suggestions are made to assist in setting up and running a corporate waste minimization program.

Make Waste Minimization Part of the Company's "Corporate Culture"

Production people are usually evaluated on whether they meet a production goal such as product quality or quantity. If the company is to encourage workers to institute and sustain waste reduction changes as part of their daily duties, the workers need to realize that minimization is an important goal of the company. The commitment has to be real, has to affect production decisions, and has to be a part of job performance evaluations for the individual workers. The message needs to come from the top, not just from the appointed environmental "nag."

Establish the Position of Waste Minimization Officer for the Company

The waste minimization officer should report to someone who has responsibility for both production and environmental compliance. When conflicts arise between the two, the company should ensure that environmental effects are deemed as important as production.

Inventory the Waste Sources by Individual Production Units

To the extent feasible, waste production should be measured and listed by individual production units. These details allow the company to evaluate individual processes to determine where waste minimization efforts should be concentrated. Waste production data are necessary to evaluate the benefits of making any changes.

Rank and Target Waste Sources by Cost of Disposal

The company should use waste production data to identify production units with either exceptionally high or exceptionally low waste production rates. To compare individual facilities, a company should relate this data to production volumes to determine if differences are real or just the result of varied rates of production. This process allows the company to invest in waste minimization efforts where they have the potential for being most effective, focusing minimization efforts on waste sources that are the most costly.

It has been proposed that priorities should be placed on waste minimization efforts for wastes having a relatively high degree of risk to human health or environment. In practice, this approach has the inherent difficulty of requiring quantification of something that is basically subjective (i.e., the degree of risk associated with disposal of a particular waste type). In fact, waste disposal costs are roughly proportional to the hazards associated with their disposal. So, simply ranking wastes by cost of disposal is a reasonable method for including risk in targeting wastes for minimization efforts.

Example. The selected waste sources shown in Table 2.2 were ranked by cost of disposal. The ammoniated citric acid and phosphoric acid were generated in the process of cleaning liquid oxygen pipelines. The material was classified as hazardous because of corrosivity (pH). By neutralization and mixing, this waste product can be converted into a liquid fertilizer for use at the facility. It was also recommended that waste battery acid be considered for use as a pH adjustment chemical in the waste treatment plant. In contrast to the amount of sludge produced in similar facilities, the sludge produced in treating metal plating wastes was negligible because of low production rates and the platers' established practice of rinsing parts over the plating tank with a hose prior to placing the parts in the flowing rinse. In this instance, no action was required, other than to commend the platers and encourage them to continue their exemplary practices.

Table 2.2. Annual Waste Generation Rates for Example Laboratory

Description	ID No.	Annual Generation Rate	Units	Unit Cost	Total Disposal Cost
Ammoniated citric acid	AB22A03	80,000	Gal	$ 2	$160,000
Phosphoric acid	AB22A06	30,000	Gal	1	30,000
Paint, miscellaneous	AB22A13	54	Drums	500	27,000
Aerosol cans	CN74A01	24	Drums	500	12,000
Trichloroethylene	AB22A02	40	Drums	259	10,360
Battery acid	AB21C02	12	Drums	89	1,068
Plating sludge	AB31H01	8	Drums	63	504

Publicize Waste Production Costs

Frequently, finding out the costs of individual waste disposal sufficiently motivates a production supervisor to make a change.

Example. At an aerospace manufacturing facility, a shop foreman was told that it was costing $600 per drum to dispose of shop rags that were contaminated with chlorinated solvents. He switched to nonchlorinated solvents for all but critical uses, allowing the rags to be cleaned and sold to a recycler.

Charge Production Units for Waste Disposal Costs

Production budgets should include the costs of hazardous waste disposal so that the costs can be used in production decisions. The budget should also include a contingency fund for potential liabilities.

Provide Capital Investment Budget for Project Funds

Combining production benefits with waste reduction savings can often justify the costs of a project that cannot be justified on either basis alone. Providing funding for (and help in justifying) waste minimization projects separate from the normal budget for capital improvements can expedite implementation of waste minimization projects.

Example. The PMB paint stripping facility at Hill Air Force Base was funded by a Productivity Enhancement Capital Investment fund that allowed construction in less than 2 years, rather than the usual 5-year cycle for military construction budgets.

Reward Successes; Learn from Failures; Publicize Both

Recognition is a powerful motivator. Rewarding successes is a means of affirming an innovator's decision to do something different; rewards encourage others to put in extra effort to reduce waste so that they can also be recognized and rewarded. At the same time, failures should be advertised (anonymously) so that others can avoid the same mistakes.

Provide Funds for Technology Transfer and Training

A company should encourage the transfer of successful waste minimization methods within its organization. Hands-on workshops or informal training programs at successful production units are particularly effective.

TIPS FOR IMPLEMENTING INDIVIDUAL PROJECTS

Inform Production Unit Personnel of Costs of Waste Disposal

Frequently, personnel do not realize the true cost of waste disposal. Just knowing the costs can be sufficient motivation for production people to make changes that will eliminate or reduce waste disposal costs.

Identify a Successful Production Unit as an Example

Methods used at low waste-producing facilities can supply ideas for changes at problem facilities. An interchange of ideas between successful and problem units should be encouraged.

Brainstorm Solutions for the Targeted Production Unit

Encourage environmental and production personnel (the supervisor and individual production workers) to develop potential solutions to waste generation problems identified in the targeted production unit.

Evaluate Ideas

The company should base its decisions on projected waste reduction, production benefits, and practicality. Factors such as low maintenance requirements and simplicity are important considerations.

Identify and Support a Champion

Ideally, a Champion should be the individual who originated the idea and who has extensive experience with the process. This person should be given the responsibility and the authority to make the change work.

Fund a Demonstration Project

A demonstration project will help to determine if the change is practical for an entire production unit. The demonstration should include testing to determine the actual reduction of wastes, effects on production costs, ease of operation and maintenance, and effects on product quality.

Evaluate the Demonstration Project

An evaluation of the project should be based on monitored cost savings, practicality, and effects on product quality.

Provide Training for Process Change

It is important to train extra operators for any new equipment. When necessary, these people can provide backup operation.

SOURCES OF ADDITIONAL INFORMATION

State Waste Minimization Programs

A number of states have established information transfer programs to assist industries in identifying methods of reducing or recycling hazardous wastes. Information is transferred by means of studies, conferences, workshops, telephone hotlines, information clearinghouses, and training programs.

Technical assistance programs (TAPs) provide individualized help to waste generators. Assistance is usually provided in the form of onsite consultation. Assistance may consist of a review of the facilities operation and suggestions of specific modifications. Because qualified experts are required to ensure that the appraisals and advice are sound and accurate, TAPs are expensive to establish and maintain. TAPs have generally been most helpful to small quantity generators, who lack the funds or expertise to investigate waste minimization on their own.

A listing of state waste minimization programs is provided as Table 2.3.

Table 2.3.   State Waste Minimization Programs

Alabama	Hazardous Material Management and Resource Recovery Program University of Alabama P.O. Box 6373 Tuscaloosa, AL 35487-6373 (205) 348-8401
Alaska	Alaska Health Project Waste Reduction Assistance Program 431 West Seventh Avenue Anchorage, AK 99501 (907) 276-2864
Arkansas	Arkansas Industrial Development Commission One State Capitol Mall Little Rock, AR 72201 (501) 371-1370
California	Alternative Technology Section Toxic Substances Control Division California Department of Health Services 714/744 P Street Sacramento, CA 94234-7320 (916) 322-5347
Connecticut	Connecticut Hazardous Waste Management Service Suite 360 900 Asylum Avenue Hartford, CT 06105 (203) 244-2007

Table 2.3. Continued

	Connecticut Department of Economic Development 210 Washington Street Hartford, CT 06106 (203) 566-7196
Georgia	Hazardous Waste Technical Assistance Program Georgia Institute of Technology Georgia Technical Research Institute Environmental Health and Safety Division O'Keefe Building, Room 027 Atlanta, GA 30332 (404) 894-3806
	Environmental Protection Division Georgia Department of Natural Resources Floyd Towers East, Suite 1154 205 Butler Street Atlanta, GA 30334 (404) 656-2833
Illinois	Hazardous Waste Research and Information Center Illinois Department of Energy and Natural Resources 1808 Woodfield Drive Savoy, IL 61874 (217) 333-8940
	Industrial Waste Elimination Research Center Pritzker Department of Environmental Engineering Alumni Building, Room 102 Illinois Institute of Technology 3200 South Federal Street Chicago, IL 60616 (312) 567-3535
Indiana	Environmental Management and Education Program Young Graduate House, Room 120 Purdue University West Lafayette, In 47907 (317) 494-5036
	Indiana Department of Environmental Management Office of Technical Assistance P.O. Box 6015 105 South Meridian Street Indianapolis, IN 46206-6015 (317) 232-8172

Table 2.3. Continued

Iowa	Iowa Department of Natural Resources Air Quality and Solid Waste Protection Bureau Wallace State Office Building 900 East Grand Avenue Des Moines, IA 50319-0034 (515) 281-8690
	Center for Industrial Research and Service 205 Engineering Annex Iowa State University Ames, IA 50011 (515) 294-3420
Kansas	Bureau of Waste Management Department of Health and Environment Forbes Field, Building 730 Topeka, KS 66620 (913) 296-1607
Kentucky	Division of Waste Management Natural Resources and Environmental Protection Cabinet 18 Reilly Road Frankfort, KY 40601 (502) 564-6716
Maryland	Maryland Hazardous Waste Facilities Siting Board 60 West Street, Suite 200A Annapolis, MD 21401 (301) 974-3432
	Maryland Environmental Service 2020 Industrial Drive Annapolis, MD 21401 (301) 269-3291 (800) 492-9188 (in Maryland)
Massachusetts	Office of Safe Waste Management Department of Environmental Management 100 Cambridge Street, Room 1904 Boston, MA 02202 (617) 727-3260
	Source Reduction Program Massachusetts Department of Environmental Quality Engineering 1 Winter Street Boston, MA 02108 (617) 292-5982

Table 2.3.   Continued

Michigan	Resource Recovery Section Department of Natural Resources P.O. Box 30028 Lansing, MI 48909 (517) 373-0540
Minnesota	Minnesota Pollution Control Agency Solid and Hazardous Waste Division 520 Lafayette Road St. Paul, MN 55155 (612) 296-6300
	Minnesota Technical Assistance Program W-140 Boynton Health Service University of Minnesota Minneapolis, MN 55455 (612) 625-9677 (800) 247-0015 (in Minnesota)
	Minnesota Waste Management Board 123 Thorson Center 7323 Fifty-Eighth Avenue North Crystal, MN 55428 (612) 536-0816
Missouri	State Environmental Improvement and Energy Resources Authority P.O. Box 744 Jefferson City, MO 65102 (314) 751-4919
New Jersey	New Jersey Hazardous Waste Facilities Siting Commission Room 614 28 West State Street Trenton, NJ 08608 (609) 292-1459 or 292-1026
	Hazardous Waste Advisement Program Bureau of Regulation and Classification New Jersey Department of Environmental Protection 401 East State Street Trenton, NJ 08625 (609) 292-8341
	Risk Reduction Unit Office of Science and Research New Jersey Department of Environmental Protection 40 East State Street Trenton, NJ 08625 (609) 633-1378

Table 2.3. Continued

New York   New York State Environmental Facilities Corporation
                        50 Wolf Road
                        Albany, NY 12205
                        (518) 456-4139

                        Division of Solid and Hazardous Waste
                        New York Department of Environmental Conservation
                        50 Wolf Road
                        Albany, NY 12233
                        (518) 457-3273

North Carolina   Pollution Prevention Pays Program
                        Department of Natural Resources and Community
                           Development
                        P.O. Box 27687
                        512 North Salisbury Street
                        Raleigh, NC 27611
                        (919) 733-7015

                        Governor's Waste Management Board
                        325 North Salisbury Street
                        Raleigh, NC 27611
                        (919) 733-9020

                        Technical Assistance Unit
                        Solid and Hazardous Waste Management Branch
                        North Carolina Department of Human Resources
                        P.O. Box 2091
                        306 North Wilmington Street
                        Raleigh, NC 27602
                        (919) 733-2178

Ohio   Division of Solid and Hazardous Waste Management
                        Ohio Environmental Protection Agency
                        P.O. Box 1049
                        1800 WaterMark Drive
                        Columbus, OH 43266-1049
                        (614) 481-7200

                        Ohio Technology Transfer Organization
                        Suite 200
                        65 East State Street
                        Columbus, OH 43266-0330
                        (614) 466-4286

Oklahoma   Industrial Waste Elimination Program
                        Oklahoma State Department of Health
                        P.O. Box 53551
                        Oklahoma City, OK 73152
                        (405) 271-7353

Table 2.3. Continued

Oregon	Oregon Hazardous Waste Reduction Program Department of Environmental Quality 811 Southwest Sixth Avenue Portland, OR 97204 (503) 229-5913
Pennsylvania	Pennsylvania Technical Assistance Program 501 F. Orvis Keller Building University Park, PA 16802 (814) 865-0427
	Bureau of Waste Management Pennsylvania Department of Environmental Resources P.O. Box 2063 Fulton Building 3rd and Locust Streets Harrisburg, PA 17120 (717) 787-6239
	Center for Hazardous Materials Research 320 William Pitt Way Pittsburgh, PA 15238 (412) 826-5320
Rhode Island	Ocean State Cleanup and Recycling Program Rhode Island Department of Environmental Management 9 Hayes Street Providence, RI 02908-5003 (401) 277-3434 (800) 253-2674 (in Rhode Island)
	Center for Environmental Studies Brown University P.O. Box 1943 135 Angell Street Providence, RI 02912 (401) 863-3449
Tennessee	Center for Industrial Services Suite 401 226 Capitol Boulevard Building University of Tennessee Nashville, TN 37219-1804 (615) 242-2456
Virginia	Office of Policy and Planning Virginia Department of Waste Management 11th Floor, Monroe Building 101 North 14th Street Richmond, VA 23219 (804) 225-2667

Table 2.3.   Continued

Washington	Hazardous Waste Section Mail Stop PV-11 Washington Department of Ecology Olympia, WA 98504-8711 (206) 459-6322
Wisconsin	Bureau of Solid Waste Management Wisconsin Department of Natural Resources P.O. Box 7921 101 South Webster Street Madison, WI 53707 (608) 266-2699
Wyoming	Solid Waste Management Program Wyoming Department of Environmental Quality Herschler Building, 4th Floor, West Wing 122 West 25th Street Cheyenne, WY 82002 (307) 777-7752

League of Women Voters

The League of Women Voters has been involved in hazardous waste management, and minimization has been a major focus of their program. Recycling and source reduction receives substantial coverage in *The Hazardous Waste Exchange*, a quarterly newsletter published by the League.

Contact: Ms. Sharon Lloyd
          Project Manager
          Citizen Involvement on Hazardous Waste Management
          League of Women Voters
          1730 M Street, NW
          Washington, DC 20036
          (202) 429-1965

Pollution Probe Foundation

The Pollution Probe Foundation is a public interest group that has been working to improve the Canadian environment. The group had concentrated on waste minimization as their principal means of attacking hazardous waste problems. In 1982, the group published the book *Profit From Pollution Prevention*, a guide to waste reduction and recycling. The group is active in conferences and symposiums.

Contact: Ms. Monica E. Campbell
Pollution Probe Foundation
12 Madison Avenue
Toronto, Ontario M5R 2S1
(416) 978-6155

Hazardous & Solid Waste Minimization & Recycling Report

Government Institutes, Inc., publishes *Hazardous & Solid Waste Minimization & Recycling Report* as a source of information for industrial waste generators, engineering companies, equipment vendors, and government regulators. The newsletter provides a monthly forum for industry practitioners to share ideas—both generic and industry-specific—for developing and implementing waste minimization, recycling, and exchange strategies.

The newsletter provides at least one case study each issue on a waste minimization technique or program that has been used successfully in industry. Reports on waste minimization technical advances, including federal research and development efforts, state-funded demonstration efforts, along with updates and analyses of the latest economic and regulatory incentives for waste minimization, are included.

INFORM

INFORM is a research organization dedicated to issues dealing with problems concerning land use, water quality and conservation, energy technologies, pollution and toxic waste management, and safety and health in the workplace. INFORM has completed a study on waste minimization, which evaluated waste management practices of 29 chemical manufacturing companies. The results of this study were published in the book, *Cutting Chemical Wastes* (1985).

Contact: Mr. Dave Sorokin
INFORM
381 Park Avenue South
New York, New York 10016
(212) 689-4040

Additional Useful Groups and Trade Associations

American Electroplaters' Society
1201 Louisiana Avenue
Winter Park, Florida 32789
(305) 647-1197

Association of Petroleum Re-Refiners
Suite 913
2025 Pennsylvania Avenue
Washington, DC 20006
(202) 833-2694

National Association of Solvent Recyclers
1406 Third National Building
Dayton, Ohio 45402
(513) 223-0419

National Center for Resource Recovery
1211 Connecticut Avenue, NW
Washington, DC 20036
(202) 223-6154

National Paint and Coatings Association
1500 Rhode Island Avenue
Washington, DC 20005
(202) 462-6272

GENERAL REFERENCES

Huisingh, D., L. Martin, H. Hilger, and N. Seldman. "Proven Profits From Pollution Prevention: Case Studies in Resource Conservation and Waste Reduction," Institute for Local Self-Reliance, Washington, D.C. (1986).

Kohl, J., P. Moses, and B. Triplett. "Managing and Recycling Solvents," Industrial Extension Service, School of Engineering, North Carolina State University, Raleigh, NC (December 1984).

Kohl, J., and B. Triplett. "Managing and Minimizing Hazardous Waste Metal Sludges," Industrial Extension Service, School of Engineering, North Carolina State University, Raleigh, NC (December 1984).

Roberts, R. M., J. L. Koff, and L. A. Karr. "Hazardous Waste Minimization Initiation Decision Report," Technical Memorandum TM 71-86-03, Naval Civil Engineering Laboratory, Port Hueneme, California (January 1988).

"Waste Minimization—Issues and Options," U.S. Environmental Protection Agency, Office of Solid Waste and Emergency Response, EPA/530-SW-86-041, Washington, D.C. (October 1986).

CHAPTER 3

Machining and Other Metalworking Operations

PROCESS DESCRIPTIONS AND SOURCES OF WASTES

In machining, a cutting tool or abrasive materials removes metal from a metal workpiece to produce a desired shape and dimension. Thus, machining is but one aspect of metalworking, which includes sawing, milling, grinding, drilling, boring, reaming, turning, stamping, forging, shaping, and heat treating.

Most metalworking operations involve high pressure, metal-on-metal moving contacts between tools and workpieces. These contacts result in high friction and generation of heat. If left uncontrolled, this heat and friction can cause excessive wear on tools and undesirable metallurgical transformations in the workpieces. To reduce these effects, metalworking fluids are circulated over working surfaces, reducing friction, cooling the tool and the workpiece, and removing metal chips from the work face. A variety of cutting oils and coolants have been developed to perform these functions. These metalworking fluids are typically stored in open reservoirs, located in either the individual machine or in a central reservoir serving a number of similar machines.

Modern metalworking fluids are of four basic types: straight oil, synthetic, semi-synthetic, and soluble oil. Table 3.1 lists the composition of the concentrates used

Table 3.1. General Classes of Metalworking (Contact) Fluids

Class	Petroleum Oil Concentration[a]	Appearance of New Mix
Synthetic	0	Transparent or opaque
Semi-synthetic	2–30%	Transparent, translucent, or opaque
Soluble oil	60–90%	Opaque
Straight oil	100%	Opaque

Source: Reference 1.
[a]Measured in concentrate prior to direct use or dilution with water to make working fluid.

33

to prepare each of these fluids and describes their appearance when mixed (all but straight oil are mixed with water to produce a working fluid).

Straight oils are used least frequently because of such past health and safety problems as fire hazards, slippery floors, and potential respiratory problems among workers who breathe oily mists. Other considerations include higher cost, poor work area appearance, and difficulties with treatment and disposal of used oils.[1]

Synthetic fluids have no oil but contain a wide range of water-soluble chemicals that provide the desired lubricating properties. These synthetics rely on water to cool and clean the work face.

The most commonly used metalworking fluid today is soluble oil coolant, since it is of comparable quality and applicability as synthetic fluids and is less expensive. Soluble oils are prepared by mixing a concentrate with water in a 1:15 to 1:20 ratio to produce a working fluid with a water content of 90% to 98%.

The use of water as the primary ingredient in soluble oils accounts for their costing less than straight oils. A disadvantage of using soluble oils, however, is that they require more maintenance than the other fluids. For example, the concentrate must be measured and adjusted daily to ensure the proper oil-to-water ratio.

In addition to these contact operations, oils are also used in metalworking operations for noncontact purposes such as transferring energy hydraulically and lubricating gear boxes and moving parts in metalworking machines. These hydraulic fluids or lubricating oils and greases are contained within enclosed reservoirs in the individual machines. Since these fluids do not come in contact with the workpiece, they are less prone to contamination than "contact" fluids such as cutting oils or coolants.

Straight oils and greases are typically used as hydraulic fluids and for machine lubrication. Soluble oils could also be used, but equipment manufacturers usually specify that certain grades or types of petroleum oils be used to ensure optimal efficiency and minimize wear, and their warranties generally depend on following these recommendations. As a result, machine owners typically are wary of using hydraulic fluids or lubricating oils other than those specified for hydraulic, gear box, or similar applications. In addition, machine owners may be reluctant to change from something that works.

When metalworking fluids no longer meet performance requirements, they are removed from their reservoirs and disposed of. The degradation of the fluids is caused by a number of factors: metal particles or shavings (swarf), grease, tramp oil, and dirt accumulate in the coolants or contact oils. In addition, heat can cause loss of water and depletion or breakdown of additives. Also, bacteria can grow in the fluid and cause it to degrade. Leaks in hydraulic seals can result in contamination of the metalworking fluids with incompatible hydraulic and/or lubricating oils.

Noncontact oils also require replacement or restoration, although less frequently than contact fluids. While manufacturers' recommendations need to be considered, frequency of fluids replacement should also be based on operating conditions. Typical oil replacement requirements for representative machine tools are provided in Table 3.2.

Table 3.2.   Representative Waste Oil Generation

Example Operation	Typical Machine	Oil Use and Purpose	Typical Waste Oil Generation
Forging	5,000-ton hydraulic press	Soluble oil—hydraulics	1,000 gal/yr
		Soluble oil—coolant	500 gal/yr
		Lubricating oil—moving parts	150 gal/yr
Grinding	Vertical grinder	Hydraulic oil—hydraulics	70 gal/yr
		Gear oil—gear box	70 gal/yr
Metal cutting	Orbital saw (abrasive blade)	Soluble oil—coolant liquid; sludge	7,500 gal/yr 10 yd³ (dry)
		Hydraulic oil— hydraulics, gear box	100 gal/yr
		Oil—metal parts lubricating	100 gal/yr
	Horizontal lathe	Soluble oil—coolant	750 gal/yr
		Straight oil—coolant, gearbox	20 gal/yr
		Hydraulic oil—hydraulics	75 gal/yr
Die sinking/ milling	Milling machine, vertical lathe	Soluble oil—coolant	2,500 gal/yr
		Hydraulic oil—hydraulics	50 gal/yr
		Penetrating oil—cutting operation	200 gal/yr
		Gear oil—gear box	15 gal/yr
Stamping	Stamping	Soluble oil—coolant	500 gal/yr
		Hydraulic oil—hydraulics	25 gal/yr

ENVIRONMENTAL REGULATIONS

On November 29, 1985, the U.S. Environmental Protection Agency (EPA) proposed to regulate used oil as hazardous waste.[2] However, the responses to this proposal and the demonstrated ability of the oil recycling industry to safely recycle oils or recover energy by burning them led the EPA to conclude that used oils are not hazardous unless contaminated with regulated hazardous waste or containing hazardous constituents. Therefore, metalworking fluids can often be classified as "waste oil," rather than hazardous waste under federal and state laws.

Wastewaters discharged to a publicly owned treatment works must contain less than 100 ppm of total oil. Oily wastewaters discharged to streams or navigable waters are subject to similar restrictions. Stormwater containing oil has also come under regulation in recent years.

METHODS TO REDUCE METALWORKING WASTES

The volume of waste metalworking fluids can be reduced through operational changes and by investing in equipment to recondition and recycle waste fluids. Since equipment to recondition waste oils can be expensive, methods that do not require capital expenditure should be explored first. A program to reduce the disposal of waste fluids should, therefore, proceed in four steps:

1. inventory materials and procedures

2. substitute materials or change processes

3. attempt regulatory exemption

4. reuse or recycle fluids

INVENTORY MATERIALS AND PROCEDURES

Preparing a current inventory of the metalworking, hydraulic, and lubricating fluids being used is the first step in a waste reduction program. This step is undertaken to confirm that the materials and procedures being used are in accordance with plant engineering policy or manufacturer's specifications. One goal of this evaluation should be to reduce the number and quantities of fluids used. Consolidating fluids can significantly reduce the volume of waste produced and simplify recycling or disposing of the wastes remaining.

Specifically, the quantity, type, and intended performance of each metalworking fluid, along with reasons and locations for its use, should be determined. It is not unusual to find dozens of different oils and other metalworking fluids being used within a single plant or on one factory floor. These fluids are selected by plant personnel based on preference, cost, quality, performance, or longevity. Plant personnel making these choices may include purchasing agents, process engineers, shift supervisors, maintenance staff, and production managers.

A complete inventory of oil type and practice is recommended for each machine. If not already developed, a computerized database can be compiled to include fluid used, manufacturer, quantity used, frequency of purchase, and cost. This inventory provides for objectively analyzing current operations, comparing costs, and ranking areas of concern. As part of this inventory, staff members who are involved in decisions to use certain fluids and who are most familiar with their routine effectiveness should be interviewed. Staff knowledge and training should influence later decisions involving process changes.

An inventory of waste oils generated from each machine should also be prepared. An important part of the inventory should be a laboratory characterization of the waste fluids. Work history or suspected composition of a waste oil is no substitute for an accurate analysis. This information is also needed for proper regulatory classification and disposal of waste oils. Additionally, it provides a basis for further waste minimization efforts.

SUBSTITUTE MATERIALS OR CHANGE PROCESSES

Fluids Selection

After preparing an inventory of existing metalworking fluids and waste production, an evaluation is needed to determine if the number and volume of different fluids being used can be reduced. Consolidating fluids can result in a significant reduction of waste produced. Further reduction can result if a new fluid is selected that produces less waste than current ones being used.

It has been recommended that the following performance and waste treatment factors be considered in selecting or reevaluating metalworking fluids.[1] These factors also apply to noncontact hydraulic or lubricating fluids.

- Performance—Ensure the fluid meets production and maintenance objectives by requiring laboratory screening tests.

- Health and Safety—Maintain low risk to operators by evaluating material safety data sheets and other information supplied by the manufacturer.

- Waste Treatment—Require screening tests, field tests, or supplier assistance to define treatment requirements, if any.

- Quality Standard of Fluid—Ask for evidence that the supplier can provide a consistent product quality.

- Technical Service Support—Request the suppliers' assistance in providing internal control and troubleshooting of degradation problems with fluids.

- Cost—Consider the overall or life-cycle cost of a fluid, not just its purchase price. Include labor for mixing, transporting, and controlling the fluids, machine downtime for adjusting and disposing of fluid, and waste treatment and disposal. Also, consider effects on operator safety and health.

- Delivery—Have supplier provide ''just in time'' delivery to minimize deterioration of fluids during storage.

A review of each fluid application must be undertaken to determine specific fluid requirements. The following factors should be considered:[1]

- type of operation

- central reservoir system or sumps on individual machines

- tooling and setup requirements or physical constraints

- type of metals being worked

- quality of water used for fluid mixing and makeup

- production part requirements (e.g., tolerances, finish, rust protection)

- machine maintenance (e.g., lubrication, seals, paint, cleanliness of work area)

Laboratory or screening tests and vendor-supplied information are useful, but the best information is obtained by testing on individual production machines. Production management, maintenance (machine overhaul and lubrication), purchasing, laboratory, environmental, and health and safety personnel should assist in the fluid selection process, but the ultimate responsibility should rest with one person or group.

Case Study 3.1: Purchase Problems Negate Coolant Recycle Program

A machine shop at the Pensacola, Florida, Naval Air Station Rework Facility operates approximately 75 drilling, grinding, milling, and lathing machines to recondition Navy A-4 jet aircraft and H-3 and H-53 helicopters. The machine shop uses an emulsified coolant, diluted to 4% concentration with deionized water. A centrifuge was purchased to remove tramp oil from the coolant, which was then being changed approximately every two to four weeks. With the new centrifuge, new coolant purchase was reduced considerably, and coolants could be used for four to six times the former life. Machine tool life was also extended.

Unfortunately, the centrifuge could be used only sporadically because the unit is effective for treating water-soluble coolants only, and the new coolants, supplied by low-bid purchase, were frequently oil-soluble. Even the same types of coolant, when supplied by different manufacturers, were found to be incompatible, so that all spent coolants had to be disposed of when the new material was introduced into the system. An inventory program, followed by reducing the number and standardizing the type of coolants, would have presented a simpler oil waste treatment problem. If this program had been implemented before the centrifuge was purchased, its effectiveness would have been increased, and its payback period would have been shortened considerably.

Improve Maintenance

In the process of inventorying and reviewing usage of metalworking fluids, operational or maintenance deficiencies can be identified. These deficiencies can often be quickly and easily remedied with minimal labor or capital expense. The goal is to reduce contamination of fluids and extend their useful lives to reduce waste generation.

For example, machine operators can extend the useful life of soluble coolants by using high-quality makeup water. Water with greater than 100 ppm of dissolved solids can contribute to corrosion because the solids may react with chlorides and sulfates, enhance bacterial growth (sulfates are biochemically reduced to hydrogen sulfide, causing a rotten egg odor), create foaming problems, increase concentrate use, and affect product quality.[1]

Additional steps can be taken to maintain coolant quality. An individual responsible for the metalworking fluids maintenance should perform the following tasks daily:

- Add makeup water to replace soluble coolant losses (5% to 15%) resulting from evaporation.

- Check the fluid pH for signs of bacteria buildup.

- Remove swarf and dirt from work area.

This individual should also routinely:

- Enforce good housekeeping practices to prevent or minimize contamination of the fluids by oil, dirt, or metal.

- See that machines are overhauled periodically, paying particular attention to tightening hydraulic, lubricating, and gear oil seals and keeping reservoirs tightly covered.

- Discourage operators from disposing of rags, paper, or other debris in coolant reservoirs or sumps.

- Control the addition of concentrate and makeup water (machine operators tend to overconcentrate or overdilute).

- Prevent the use of bactericides or other chemicals that can cause health and safety problems or can result in spent fluids that are unnecessarily toxic.

Proper machine maintenance is particularly critical to keeping waste fluid volumes low. A leaking hydraulic fluid reservoir, for example, can generate 3 to 10 times the waste volumes considered to be normal (Table 3.2). It is not uncommon for 5% to 10% of the hydraulic mechanisms in a high-production machine shop to leak because production is emphasized over maintenance.

Daily checks of soluble coolant with a refractometer or by titration are warranted. These tests should replace the operators' test of "good feel" to the hands. Avoiding physical contact is particularly important, since many modern coolants contain bactericides that are irritating to the skin. The pH can be measured with litmus paper or a pH meter (recently developed, inexpensive, compact, battery-operated pH meters allow testing in seconds).

Other measures can be taken to reduce the complexity of coolant management. Again, the primary goal should be to reduce the number of coolants used as much as possible. One is ideal. Coolant management is simplified if a central reservoir is used for all machines in a particular department or section of the plant floor, instead of using individual reservoirs for each machine.[3]

ATTEMPT REGULATORY EXEMPTION

Waste oils are exempt from state and federal hazardous waste regulations, provided they have not been contaminated by being mixed with regulated wastes. Waste oils are still controlled under other regulations, such as the National Pollutant Discharge Elimination System (NPDES) permit system for direct discharges to the nation's surface waters. Maintaining the regulatory exemption will result in cost savings because the cost of treating and safely disposing of nonhazardous oily wastes and the expense of paperwork preparation are lowered. Proven methods of maintaining the regulatory exemption for used oils include:

- segregating waste metalworking fluids from hazardous constituents such as chlorinated solvents or PCBs
- clearly labeling waste storage containers
- training operators to avoid mixing hazardous wastes with exempt wastes
- reducing the quantity of hazardous waste-contaminated fluids and length of time that they are stored to avoid becoming a hazardous waste storage facility (90 days maximum under RCRA)

REUSE OR RECYCLE FLUIDS

Reconditioning waste metalworking fluids consists of removing impurities and then restoring the fluids to their original condition by adding concentrate and individual constituents such as surfactants, bactericides, emulsifiers, conditioners, antioxidants, or other chemicals that make the fluid effective in metalworking operations. Frequently, a supplier can recondition waste fluid more cost-effectively than a single shop. The supplier can arrange to pick up waste fluid when delivering new concentrates.

However, many metalworking fluids can be reused for months or even years without reconditioning. For these fluids, contaminants such as dirt, metals, or bacteria can be removed, and concentrate can be added to restore the oil to near-original condition. Soluble oils usually require the addition of water. Noncontact fluid usually requires only removal of contaminants. The straight oils generally require reconditioning much sooner than soluble oil or synthetic fluids because of more rapid depletion of essential constituents.

Contaminants such as other oils, dirt, metals, and bacteria can be removed through a variety of recycling processes. Recycling and treatment processes available for individual machine sumps or central reservoirs are presented in Table 3.3. Some of these processes, notably ultrafiltration (UF), are waste treatment methods, not recycling processes, since the waste oil volume is concentrated to facilitate disposal rather than reuse.

Table 3.3. Recycling Systems for Metalworking Fluids[a]

Recycling Process	Removes		
	Oil	Dirt & Metals	Bacteria
Settling/drag-out		X	
Cartridge filtration		X	
Basket strainers		X	
Gravity settling	X	X	
Cyclone separator		X	
Centrifuges	X	X	
Magnetic separator		X	
Pasteurization/ distillation			X
Ultrafiltration		X	X

[a]Adapted from Reference 1.

Selecting the most applicable recycling system for a particular plant depends on several site-specific factors. Chief among these are the following:[1]

- favorable economics, considering capital, operation and maintenance costs, and savings from reduced raw material purchase and waste disposal

- equipment effectiveness at removing contaminants such as dirt, oil, and bacteria

- availability of an automatic makeup system for reconstituting the fluid

- operational simplicity and maintenance requirements

- requirement for floor space

- quality and durability of construction and warranty protection

- availability of spare parts and services from the manufacturer

Recycling or treatment equipment can be supplied in capacities to match the requirements of individual machine sumps, although it is normally more economical to service a central reservoir system.

Cleanup of contaminated metalworking fluids can involve physical separation processes (settling, flotation, straining, filtration, centrifugation, and ultrafiltration), as well as thermal processes (pasteurization and distillation). Each of these processes is described briefly. A more detailed description of these processes is described in other literature.[4-13]

The physical removal processes produce a cleaner fluid, removing finer solids, at the expense of increased equipment cost as well as head loss, which must be supplied by pumping. For example, simple settling tanks or oil/water separators can separate swarf and unemulsified oils from coolant; basket strainers can remove coarse solids from a recirculating fluid; magnetic and cyclone-type separators can remove

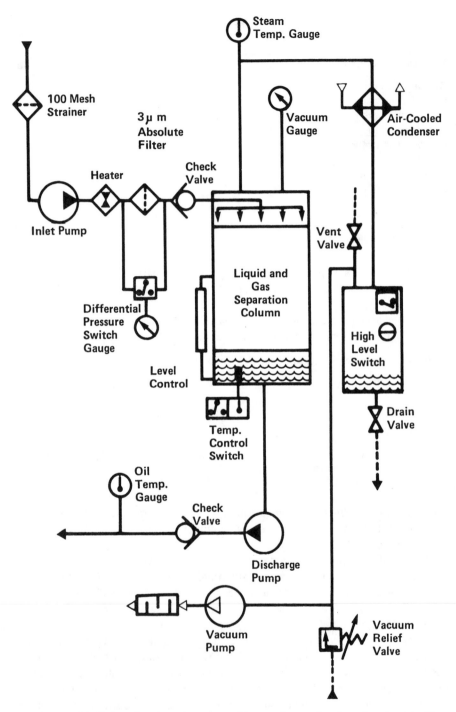

Figure 3.1. Distillation oil reclamation system (Source: Aquanetics, Inc., Farmingdale, New York).

ferrous and nonferrous metal fines, respectively; and cartridge filters can remove finer solids and metal fines.

Pasteurization and distillation involve heating contaminated fluids to between 140° and 250°F. Higher temperatures are to be avoided to prevent degradation of the fluid. Pasteurization is used to kill bacteria. Distillation (Figure 3.1) also kills bacteria, but in addition, it removes water from the fluid or oil. While distillation is commonly employed for oil/water mixtures, pasteurization is generally used for straight oil wastes.

Centrifuges employ induced high gravitational forces to separate solid/liquid or liquid/liquid mixtures that would take too long to separate under normal gravitational conditions.

To operate successfully, centrifuges usually require that the two materials to be separated have differences in specific gravity of at least 2% to 3%.

UF (Figure 3.2) employs membranes that have pores small enough to remove bacteria, microscopic solids, and oils. UF provides excellent separation (90% to 98% efficiency) of contaminants from a metalworking fluid. UF is frequently used as an end-of-pipe treatment process for oily wastes. Wastes containing from 0.1% to 10% oil, with temperatures up to 150°F and having a wide pH range, are suited to UF. A concentrate of 40% to 60% oil and a permeate of 50 ppm oil can be achieved. Pretreatment by filtration and pasteurization is normally recommended to keep the membranes from becoming plugged or fouled with dirt and bacteria.

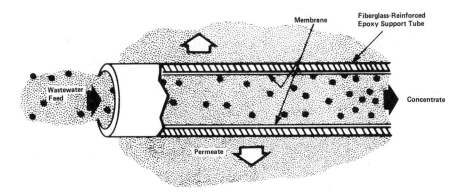

Figure 3.2. Tubular membrane ultrafiltration (Source: Membrane Systems, Inc., Wilmington, Massachusetts).

Metalworking fluid recycling systems can either use a single separation process or can employ several processes operated in series. Simple settling or magnetic separation may be sufficient to remove metal turnings from a coolant, but removal of dirt or fines can require filtration. Separation of tramp oil or bacteria may require distillation or UF. Typically, a combination of separation processes is used. A complete coolant recycle system, employing a settling tank, strainer, cyclone, pasteurizer, and centrifuge, is shown as Figure 3.3.

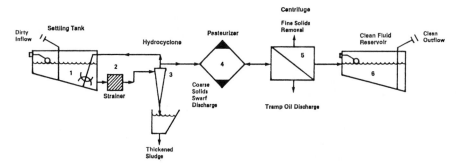

Figure 3.3. A coolant reclamation system using pasteurization and centrifugation (Source: Sanborn Inc., Wrentham, Massachusetts).

Typical capital, operation and maintenance costs, and payback periods for individual components of a fluid recycling system are listed in Table 3.4. The annual volumes shown represent the upper and lower extremes for waste oil generation from small- to medium-size metalworking operations. From these ranges, it is possible to compare costs of the different recycling equipment.

A rule of thumb in waste oil recycling is that operations generating more than 25,000 gallons per year (gal/yr) can best afford the time and staff for proper operation and maintenance of recycling equipment. Most manufacturers do not offer systems sized for less than 10,000 gal/yr, although often one unit can accommodate both ranges shown in Table 3.4. As can be seen, the simpler techniques are less capital intensive and have shorter payback periods, but the more sophisticated equipment merits consideration.

Safety-Kleen, a recycling and waste disposal company, is currently testing a mobile service unit for those machine shops that produce such low volumes of used coolants that it is uneconomical to purchase and operate recycling equipment (Figure 3.4). Waste coolant is placed into the portable unit and recycled onsite, and the user is charged per gallon of recovered oil.

In evaluating which system would be most appropriate, the machine tool operator or owner will first need to characterize the oil and its intended application. Payback periods of one to two years or less are typical and usually warrant introduction of recycling systems into plants where further waste minimization is desired. Perhaps the most important benefit of waste minimization for the plant, and ultimately for the environment, is that less waste oil requires treatment or disposal offsite. When waste oil is disposed of offsite, the owner retains responsibility under federal law(s) and is liable for environmental damage caused by improper or inadequate actions by a licensed treatment, storage, or disposal facility.

Table 3.4. Typical Waste Minimization Systems and Costs

Waste Minimization System	25,000 gal/yr			100,000 gal/yr		
	Capital Cost, $	O&M Cost ¢/Gal Processed	Payback Period, Months	Capital Cost, $	O&M Cost ¢/Gal Processed	Payback Period Months
Cartridge filters	300–1,000	3–8	<6	600–2,000	2–6	<6
Solids separators	2,000–5,000	4–10	<6	5,000–8,000	3–7	<6
Centrifuge	10,000–20,000	5–15	<24	20,000–30,000	3–12	<18
Distillation	10,000–20,000	5–15	<24	25,000–35,000	3–12	<18
Ultrafiltration	10,000–20,000	15–30	<24–36	20,000–30,000	7–15	<18

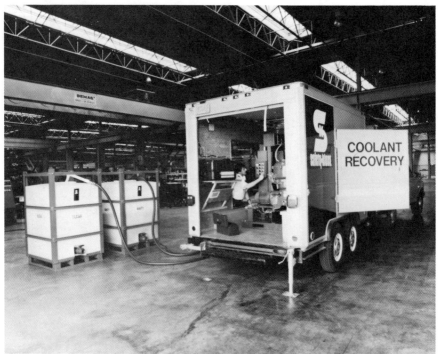

Figure 3.4.   A mobile unit used to clean metalworking fluids at a customer's plant. (Photo courtesy of Safety-Kleen, Elgin, Illinois).

Case Study 3.2: Recycle of Heat-Treating Quench Oil in Upstate New York by Settling and Cartridge Filtration

Waste oil is generated in a high-volume, heat-treating operation in which heated metal parts are quenched (rapidly cooled) from temperatures as high as 1,750°F. After cooling, the parts are spray-washed with water. The combined oil/water mixture was treated in an oil/water skimmer, which separated the waste oil and sludge for disposal. This waste oil mixture contained 4% water and 1,400 ppm of suspended solids, consisting of sediment and metal fines.

Pilot testing by Hilliard, an oil reclamation system manufacturer in Elmira, New York, demonstrated that this waste oil mixture could be treated to sufficient purity that it could be reused in the quench tank. Based on this testing, a complete treatment system was installed. Treatment processes begin with batch settling in an oil/water separator for 24 hr. Following settling, water and solids are removed from the bottom of the tank, and the oil is recirculated at a rate of 210 gal/hr through a 3-μ Hilco polypropylene fiber cartridge filter and returned to the dirty oil tank. After filtration is completed, the oil is pumped at 25 gal/hr through an oil reclaimer. In the reclaimer the oil is heated under a vacuum to reduce the water content and kill bacteria and is then filtered through a 0.5-μ Hilco cartridge filter and pumped into a clean oil tank for dispensing to the quench tank.

Settling effectively reduced the water and sediment content of the oil to 0.3% and reduced suspended solids to 120 ppm. Filtration and heat treatment reduced the concentrations of these contaminants to 75 and 5 ppm, respectively. The 3,000-gal/yr waste oil recycling system paid for itself in less than 18 months, based on a virgin oil cost of $3.82/gal and an O&M cost for the recycling system of $0.29/gal.

Case Study 3.3: Recycle of Honing Oil at a Midwestern Auto Engine Plant by Filtration, Settling, and Centrifuge

An automobile manufacturer generates waste fluids in the process of machining engine bores. The final step in machining involves use of a "honing" fluid, in this case, kerosene, to achieve a clean, mirror-like finish on the engine bore surface. During subsequent engine cleaning operations, the honing oil becomes mixed with aqueous solvents, producing a mixture that is 30% water, 70% kerosene, and less than 2% solids. To be acceptable for reuse, honing fluid can contain no more than 0.5% water and no solids larger than 5 μ.

Kerosene is difficult to recycle. When new, it is lighter than and insoluble in water, although both have similar viscosities. Having a low flash point (100 to 200°F), it is combustible and therefore presents a fire hazard.

Enhanced gravity settling and low-speed (less than 2,000 G) centrifuges were used unsuccessfully to reclaim this waste honing fluid. These systems failed because of inability to meet specifications (settling) or a buildup of solids (centrifuge).

Following consultation with Sanborn, an equipment vendor in Wrentham, Massachusetts, a high-speed centrifuge was installed. Pretreatment consists of magnetic separation and filtration through a paper media roll filter, followed by gravity oil/water separation, from which the water is disposed of.

Dewatered fluid is then heated to 150°F and fed at a rate of 3 gal/min to a centrifuge in which remaining fines and dispersed and emulsified water are separated from the honing fluid. Cleaned fluid is pumped through a heat exchanger to recover energy, then into a clean tank for reuse.

This system has operated successfully for the past year, with 500 to 700 gal of honing oil recovered each week. Payback for the system, including reduced waste disposal costs, has been less than 8 months, based on a purchase and handling cost for fresh kerosene of $1.60/gal.

Case Study 3.4: Distillation Recycle of Hydraulic Oils at an Auto Engine Assembly Plant in the Midwest

An automotive manufacturer operates more than 3,000 pieces of hydraulically operated power equipment and machinery at a single large engine plant. Fluids had to be replaced because of contamination by water and dirt. Careful plant maintenance practices were employed to prevent mixing the hydraulic fluids with coolants or other fluids. Replacing straight oil hydraulic fluids at the plant cost more than $60,000 per year. Faced with this cost, the manufacturer decided to install an onsite recycling system to recover waste hydraulic fluids.

An oil recycling system manufactured by Aquanetics, Farmingdale, New York, was installed to handle all of the plant's hydraulic fluids at a central facility. Dirty hydraulic fluids are delivered to one of three 2,000-gal, steam-heated (150°F) tanks. In these tanks, water and grit are removed by gravity settling. Heated oil is then pumped through a 50-μ cartridge filter to another tank. From this tank, the oil is pumped through a 100-mesh strainer and a 3-μ cartridge filter to a vacuum distillation unit operated at a temperature of 200 to 250°F. Water and other vapors are discharged to sewers, and the cleaned oil is pumped to a 2,500-gal holding tank for cooling. The cooled oil is tested for quality and then recycled.

The resulting cleaned oil has met or exceeded specifications for fresh oil. In five years of operation, the recycling system has recovered an average of 300 gal/day of hydraulic oil, significantly reducing fresh oil purchase and waste oil disposal. The system paid for itself in less than a year.

Case Study 3.5: Ultrafiltration for Waste Oil at an Auto Body Stamping Plant in Ontario, Canada

An automobile manufacturer operates a high production metal stamping operation in Ontario to produce automobile frames, axle housings, and body extensions. A forming oil is used in the stamping process. Following stamping, the parts are cleaned with a hot (150°F) dilute (1.5%) alkaline aqueous cleaner containing emulsifying agents. After becoming contaminated with forming oil (11,000 ppm) and metal particles, this cleaner was previously disposed of without treatment in a municipal sewer.

Changing regulations prohibited continued discharge of this untreated waste to the municipal sewer. Since offsite disposal cost for this waste would have exceeded $300,000/yr (1983 Canadian dollars), the feasibility of treatment to meet sewer discharge requirements was evaluated.

Laboratory-scale treatability testing proved that conventional chemical and physical treatment methods were insufficient to achieve discharge standards. UF proved more successful and was adopted for full-scale implementation.

A treatment system was installed, consisting of a heat exchanger for cooling, a bag filter for removing large dirt and metal particles, and a UF system for separating the forming oil. A second batch UF system was used to further treat the forming oil to a concentration of 55%. Consisting of water, alkaline cleaner, and emulsifier, the permeate from the UF system, which contained less than 200 ppm of oil, was returned to the wash tank for reuse. The volume of waste discharged to the sewer has been significantly reduced and is limited to backflushing the UF membranes and cleaning the tank.

No major difficulties were encountered in startup and operation of the system. The major components of the system are:

- two Romicon UF units, one with 20 cartridges and 15 m² of total filter area and the second with 10 cartridges and 7.5 m² of filter area

- three Gould 1,800-L/min feed pumps

- one Tranter 1.9-m^2 heat exchanger

- four Gould 110-L/min transfer pumps

- two 8.5-m^3 process tanks

- two 1.5-cubed-meter tanks to store permeate for UF membrane cleaning

- automatic level controls and associated equipment to control the process

The system operates continuously, except for batch concentration of the forming oil. Economics of the system are shown in Table 3.5.

Based on these installed costs, an annual O&M cost of $30,000 for the new facility, as well as these recovered costs, the payback period was less than one year.

Table 3.5. Cost Savings Due to Installation of Ultrafiltration Oil Treatment System (Canadian Dollars, 1983)

Previous Annual Cost:	
Alkaline cleaner	$100,000
Waste oil hauling	293,000
Total	$393,000
Study, System Design, and Installation:	
Treatability studies	$ 24,000
Detailed engineering	36,000
Romicon UF system	126,000
Support equipment and installation	96,000
Total	$282,000
Annual Cost Savings:	
Alkaline cleaner (50%)	$ 50,000
Waste oil hauling (90%)	260,000
Recovered forming compound (20%)	70,000
Total	$380,000

REFERENCES

1. Dick, R. M., "Metalworking Fluid Management." Cincinnati Milacron, Cincinnati, Ohio.
2. "Proposed Used Oil Regulations; Request for Comments on Proposed Regulations," *Federal Register*, 50:49258 (Nov. 29, 1985), 50:8206 (March 10, 1986).
3. Minetola, A.J., "Bulk Lubricant Storage and Distribution—Some of the Problems," *Lubrication Engineering* 37(5):268–271 (May 1981).
4. Perry, R. H., and C. H. Chilton, *Chemical Engineer's Handbook*, 5th ed. (New York: McGraw-Hill Book Company).

5. Dick, R.M., "Ultrafiltration for Oily Wastewater Treatment," in *Proceedings of the 36th Annual American Society of Lubricating Engineers Meeting* (May 11–14, 1981). Available from the Society of Tribologists and Lubrication Engineers, Chicago, IL.

6. Hachadoorian, R.H., "Recycling Industrial Oil," *Eng. Dig.* (December, 1984).

7. Springborn, R.K., ed. "Cutting and Grinding Fluids: Selection and Application," Dearborn, MI: American Society of Tool and Manufacturing Engineers, 1967.

8. Napier, S., "Waste Treatability of Aqueous-Based Synthetic Metal-Working Fluids," *Lubrication Engineering* 41(6):361–365 (June 1985).

9. Zelnio, L.L. "Reclamation of Metalworking Fluids from Individual Machine Tools," *Lubrication Engineering* (January, 1987).

10. Brinkman, D.W. "Technologies for Re-Refining Used Lubricating Oil," *Lubrication Engineering* (May, 1987).

11. Swain, J.W., Jr. "Conservation, Recycling, and Disposal of Industrial Lubricating Fluids," *Lubrication Engineering* (September 1983).

12. Webb, H.R. "Establishing Oil Recovery and Reclamation Programs," *Lubrication Engineering* 39(10):626–630 (October 1983).

13. Davies, R., R. Curtis, and R. Laughton, "Recycling Metal Stamping Plant Wastes," *Water and Pollution Control* (September/October 1985).

GENERAL REFERENCE

Code of Federal Regulations (CFR). Section 40, Chapters 261–270: RCRA Regulations.

CHAPTER 4

Solvent Cleaning and Degreasing

DESCRIPTIONS OF SOLVENT CLEANING PROCESSES AND WASTES

Solvent cleaning and degreasing is the process of using an organic solvent to remove unwanted grease, oils, and other organic films from surfaces. The pollutants generated include (1) the liquid waste solvent and degreasing compounds containing unwanted film material and (2) the air emissions containing volatile solvents. In almost all cases, solvent cleaning and degreasing is also used to prepare surfaces for painting.

There are three distinct types of operations that fit into the category of solvent cleaning and degreasing: cold cleaning, vapor degreasing, and precision cleaning. These processes and waste sources are described below.

Cold Cleaning

Cold cleaning is the simplest, least costly, and most common type of solvent cleaning. The solvent is usually applied at ambient temperature or is heated slightly. It is applied either by brush or by dipping the items to be cleaned in a solvent dip tank. The most common solvent applied for cold cleaning is known by the proprietary commercial names of Stoddard Solvent or Varsol. This solvent is a highly flammable mineral spirit. Generally the least expensive solvent used for cleaning, it costs between $1.00 and $2.50 per gallon.[1]

Vapor Degreasing

Vapor degreasing uses nonflammable, chlorinated hydrocarbons in a vapor phase to clean metallic and other suitable surfaces. The special apparatus that provides solvent vapor consists of a tank that is one-tenth full of solvent (Figure 4.1). The solvent is heated, usually by steam coils, to its boiling point, producing solvent-saturated vapor in the upper portion of the tank. The item to be cleaned is either inserted manually or automatically into the vapor region, where hot solvent vapor immediately condenses onto the surface of the item. The condensed solvent then drips back into the liquid bath and takes with it the removed dirt and grease. The solvent vapor is usually prevented from escaping to the atmosphere by use of a

51

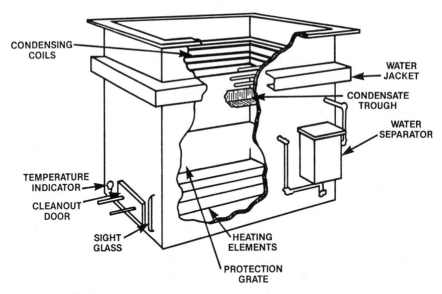

Figure 4.1.   Solvent vapor degreaser.

refrigerated section on the upper part of the tank. The solvent condenses on the walls of this section and returns to the liquid sump.

The most common solvents used in vapor degreasing are trichloroethylene; perchloroethylene; 1,1,1-trichloroethane; and methylene chloride.[2] Trichloroethylene is the most popular vapor degreasing solvent. Its relatively low boiling point (190°F) allows the use of low-pressure steam for heating and permits the handling of parts immediately after cleaning. The second most popular solvent is 1,1,1-trichloroethane, which has an even lower boiling point than trichloroethylene. However, because it is reactive with zinc and aluminum, it cannot be used to clean those materials. Perchloroethylene is used in approximately 15% of vapor degreasing applications. It is stable, has a high boiling point, and is the least aggressive solvent. Other solvents used in vapor degreasing include fluorocarbons (freon) and carbon tetrachloride.

Precision Cleaning

Precision cleaning of instruments and electronic components requires solvents of high purity, high solvency, and rapid evaporation rates. Freon compounds are customarily used for these applications.

ALTERNATIVES TO CHLORINATED SOLVENT CLEANING

Chlorinated solvents were adopted by industry because of their ability to dissolve a wide range of organic contaminants, low flammability, and high vapor pressure, which allows them to evaporate at room temperature and leave a residue-free surface. However, concerns about the toxicity and environmental effects (ozone depletion in the upper atmosphere) of continued use of chlorinated solvents have led chemical companies to search for replacement solvents.

Alternative Organic Solvents

Alternative solvents have been proposed, including those produced from citrus products. However, these products have not as yet been proven to be nontoxic nor have they reached the point of general use.

Because of the difficulty involved in disposing of chlorinated solvents, it is prudent to reconsider their use. Although flammability hazards are present when cleaning with straight hydrocarbons, the reduced costs of disposal of these nonlisted wastes can be worth the changeover from chlorinated solvents.

Alkaline Cleaners

Some companies have been successful in converting from vapor degreasers to alkaline cleaners for removal of oils and greases.

Example. The Torrington Company manufactures automobile bearings. Oils are used in stamping and quenching operations in the manufacturing of the bearings. Previously, 1,1,1-trichloroethane was used in a vapor degreaser to remove these stamping and quenching oils. When the company expanded in 1982, they switched to a hot water alkaline cleaner, using an automated parts washer manufactured by the Jenson Fabricating Engineers of Berlin, Connecticut. The custom-made washer, which included a hot air dryer, cost around $40,000 installed and was estimated to have had a payback period of one year, principally due to reduced solvent costs.[3]

High Pressure Hot Water Washers

Hot water washers can be used without solvents or detergents in applications such as cleaning oil and grease from engine compartments. Suppliers of high pressure hot water washers are listed in Table 4.1. A typical unit is shown in Figure 4.2.

Table 4.1. Suppliers of Self-Contained High Pressure Hot Water Washers

Name	Address	Phone
Hydro Systems Co.	3798 Round Bottom Road Cincinnati, OH 45244	(513) 271-8800
Sioux Steam Cleaner Co.	Sioux Plaza Beresford, SD 57004	(605) 763-2776

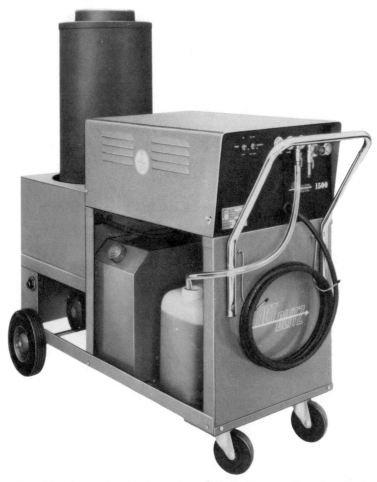

Figure 4.2. Hot water washer (Photo courtesy of Hydro Systems Company, Cincinnati, Ohio).

Case Study 4.1: High Pressure Hot Water Washers for Vehicle Maintenance Cleaning

Cleaning and maintaining military vehicles can produce a waste that is difficult to treat. At Fort Polk, Louisiana, cleaning engine compartments with solvents, steam, and detergents resulted in an emulsified, frothy waste that had a consistency like chocolate mousse. The steam and detergents tended to emulsify the oil in the water, making it impossible to separate in a simple oil/water separator. Solvents contaminated both the water and the oil, rendering both a hazardous waste. Disposal of this mixed waste cost $0.80 per gallon F.O.B. the disposal site in 1985, for a cost of $84,000 per year plus shipping.

Similar problems were encountered at Fort Lewis, Washington. At this Army post, however, the vehicle maintenance racks were equipped with high-pressure (800 psi) hot water washers. The self-contained, oil-fired, high-pressure hot water washers cost less than $2,000 each. These units supply 3.5 gal/min of hot water jets to a handheld wand that cuts the grease and oil from the engine compartments without the need for detergents or solvents. Replacing the old facilities where 30 gal/min of cold water and steam cleaners were used, the hot water washers have reduced both water usage and maintenance.

Since the oil and grease were no longer emulsified, a simple oil/water separator was sufficient to treat this wastewater. In 1984, an additional 46,000 gal of used oil was recovered and sold to a recycler for $10,800.

RECYCLING WASTE SOLVENTS

Cascade Reuse

Frequently, high-quality solvents are used once for precision cleaning and are then disposed of, either on a scheduled basis or because they do not meet their original specifications. Such solvents could be reused without treatment for applications that do not require as high a standard of purity. Using an untreated waste from an operation with high purity standards for a less demanding function without is called cascade reuse.

Example. At NASA's Marshall Space Flight Center, freon and isopropanol are used to clean high-pressure liquefied gas lines on a periodic basis. Once used, these solvents cannot be reused for their original purpose because of contamination by small amounts of water. These used solvents could be recycled for general purpose cleaning at the facility. The used solvents can be used directly, without treatment, because general cleaning does not require the high purity solvent that pipe cleaning requires.

Example. An aerospace missile manufacturer's procurement specifications require that isopropanol be provided at work stations in a specific-size squeeze bottle, with the solvent replaced on a set schedule. As a result of this requirement, unused solvent was disposed of when its expiration date was reached even though it was not contaminated. In the future, this solvent could be collected and used for nonprohibited purposes.

Distillation

Several technologies are available for cleaning solvents so that they can be reused for their original purpose. The most promising technology that can be applied to almost all industrial facilities is distillation. Solvent recovery using distillation can be implemented in five configurations: company-owned local recycling system, company-owned central recycling system, contract recycling, sale to recyclers, and manufacturer take-back. Other technologies that were identified as commercially available and that may have specific application for a particular process are centrifugation, filtration, ultrafiltration, reverse osmosis, and activated carbon adsorption. This technology relies on heating a solvent until it vaporizes, and then condensing the vapor.[4] The condensed vapor is reused. If the boiling point of the solvent is high (over 200°F), the distillation usually takes place under a vacuum to minimize the thermal decomposition of the solvent. Another technique used for solvents with a high boiling point is to inject steam into the solvent, forming an azeotropic mixture that has a lower boiling point. The condensate of water and solvent is then separated by gravity. Table 4.2 lists the solvents amenable to distillation, along with important physical parameters. There is usually a 10:1 to 15:1 volume reduction of waste to be disposed of when recycling by distillation is used.

Table 4.2.   Physical Properties of Commonly Used Solvents

Solvent	Atmospheric Boiling Pt (°F)	Azeotropic Boiling Pt (°F)	Density (lb/gal)
Aliphatic Hydrocarbon			
Hexane	157.0	142.9	5.51
Heptane	209.0	174.8	5.70
Stoddard	308–316.0	204.0	6.47
Aromatic Hydrocarbon			
Benzene	176.0	157.0	7.32
Toluene	232.0	185.0	7.20
Xylene	261–318.0	202.1	7.17
Chlorinated Hydrocarbon			
Trichloroethylene	189.0	163.8	12.2
Perchloroethylene	249.0	189.7	13.5
1,1,1-Trichloroethane	166.0	149.0	11.0
Methylene Chloride	104.0	101.2	11.07
Fluorocarbon			
Freon TF	117.6	112.0	13.06
Freon 112	199.0	166.0	13.69
Acetone	133.0	133.0	6.59
Methyl Ethyl Ketone (MEK)	175.0	164.1	6.71
Methyl Isobutyl Ketone (MIBK)	241.0	190.2	6.67

Source: DCI Corp., Indianapolis, Indiana.

Most solvents are not pure chemicals, but are formulated as mixtures of compounds. Proprietary inhibitors are added to commercial solvents to improve their properties or extend their service life. Three types of inhibitors are added to the commonly used chlorinated solvents: trichloroethylene (TCE), trichloroethane (TCA), and tetrachloroethylene or perchloroethylene (PERC). These inhibitors are classified as antioxidants, metal stabilizers, and acid acceptors.

Antioxidants form stable compounds that suppress the decomposition reaction of unsaturated solvents and slow the propagation of auto-oxidation. Metal stabilizers inhibit solvent degradation that normally occurs in the presence of metals and their chlorides (e.g., aluminum [Al] and aluminum chloride [$AlCl_3$]). The inhibitor either reacts with the metal to form an insoluble deposit or complexes the metal chloride to prevent degradation. Acid accepters are either neutral (epoxide) or slightly basic (amine) compounds that react with hydrochloric acid (HCl) produced in the breakdown of the chlorinated solvents. If left unneutralized, HCl can further degrade the solvent and corrode degreaser equipment as well as the parts being degreased.

There is often a concern that distillation will alter or fail to recover these inhibitors and, thus, that recovered solvent will not perform satisfactorily. Joshi, et al.[5] performed a study on the fate of inhibitors when subjected to repeated use and recycle using distillation. They identified inhibitor compounds used in commercial chlorinated solvent mixtures (Tables 4.3 through 4.5). In tests of solvent reclamation by distillation, they found that the inhibitors were carried over to the reclaimed solvent in excess of 65% of their concentrations in the feed spent solvents (Tables 4.6 through 4.8).

Table 4.3. Additives/Impurities Identified in Trichloroethylene

Inhibitor	Formula	MW	BP (°F)	Function
Butylene oxide	C_4H_8O	72.1	146	Acid acceptor
Ethyl acetate	$C_4H_8O_2$	88.1	171	Unknown
5,5-Dimethyl-2-hexene	C_8H_{16}	112.2	Unknown	Unknown, possibly anti-oxidant
Epichlorohydrin	C_3H_5OCl	92.5	242	Acid acceptor
N-Methylpyrrole	C_5H_7N	81.1	239	Antioxidant

Source: Adapted from Joshi.[5]

Table 4.4. Additives Identified in Tetrachloroethylene (PERC)

Inhibitor	Formula	MW	BP (°F)	Function
Cyclohexene oxide	$C_6H_{10}O$	98.2	269	Acid acceptor
Butoxymethyl oxirane	$C_7H_{14}O_2$	130.2	Unknown	Acid acceptor

Source: Adapted from Joshi.[5]

Table 4.5.   Additives Identified in 1,1,1-Trichloroethane (TCA)

Inhibitor	Formula	MW	BP (°F)	Function
N-Methoxy-methanamine	C_2H_7NO	61.1	Unknown	Acid acceptor
Formaldehyde dimethyl hydrazone	$C_3H_8N_2$	72.1	Unknown	Aluminum stabilizer
1,4-Dioxane	$C_4H_8O_2$	88.1	214	Aluminum stabilizer

Source: Adapted from Joshi.[5]

Table 4.6.   Inhibitor Concentrations of Reclaimed Trichloroethylene (TCE)

	Inhibitor Concentration (wt fraction)			
Sample	Butylene Oxide ($\times 10^3$)	Epichloro-hydrin ($\times 10^3$)	Ethyl Acetate ($\times 10^4$)	n-Methyl Pyrrole ($\times 10^4$)
New TCE	1.64	1.66	3.46	1.59
Spent TCE	0.685	1.69	2.85	2.18
TCE distillate	0.718	1.61	2.58	1.66
Carbon adsorbed TCE	0.44	1.31	2.65	0.9

Source: Adapted from Joshi.[5]

Table 4.7.   Inhibitor Concentrations of Reclaimed Tetrachloroethylene (PERC)

	Inhibitor Concentration (wt fraction)	
Sample	Cyclohexene Oxide ($\times 10^3$)	Butoxymethyl Oxirane ($\times 10^3$)
New PERC	1.06	4.26
Used PERC	0.988	7.45
PERC distillate	0.968	5.42
Carbon adsorbed PERC	0.091	5.40

Source: Adapted from Joshi.[5]

Table 4.8. Inhibitor Concentrations of Reclaimed Methylene Chloride (MC)

Sample	N-Methoxy-methanamine ($\times 10^4$)	Formaldehyde Dimethyl Hydrazone ($\times 10^3$)	1,4-Dioxane ($\times 10^3$)
New MC	8.92	5.78	17.2
Used MC	4.14	6.16	29.0
MC distillate	4.60	7.22	19.6
Carbon adsorbed MC	1.30	3.37	23.4

Source: Adapted from Joshi.[5]

Many commercially available distillation systems can distill solvent quantities ranging from 0.5 to 100 gal/hr. The smaller systems are self-contained, off-the-shelf units that can be installed in any sheltered area that has electrical power and cooling water available (Figure 4.3). The larger units are generally more complex and require the availability of steam. The capital cost is generally about $5,000, plus $1,000/gal/hr capacity. For example, a 50-gal/hr still would cost about $55,000. Generally, the payback period for purchase of a still is between 6 months and 2 years. The normal lifetime of a still is about 20 years.

Table 4.9 lists the major suppliers of self-contained solvent distillation apparatus.

The operating costs of distillation apparatus include labor, energy, cooling water, and maintenance parts. Normally, the largest component is labor. A moderately skilled operator is needed to tend the apparatus about 10% of the time during operation.

Table 4.9. Suppliers of Self-Contained Distillation Equipment

Name	Address	Phone
Baron-Blakeslee	2001 North Janice Avenue Melrose Park, IL 60160	(312) 450-3900
Corpane Industries	250 Production Court Louisville, KY 40299	(502) 491-4433
Detrex Chemical	4000 Town Center Southfield, MI 48075	(313) 358-5800
Finish Engineering	921 Greengarden Road Erie, PA 16501	(814) 455-4478
Phillips Mfg. Co.	7334 North Clark Street Chicago, IL 60626	(312) 338-6200
Recyclene Products	405 Eccles Avenue, South Cincinnati, OH 45226	(513) 321-9178

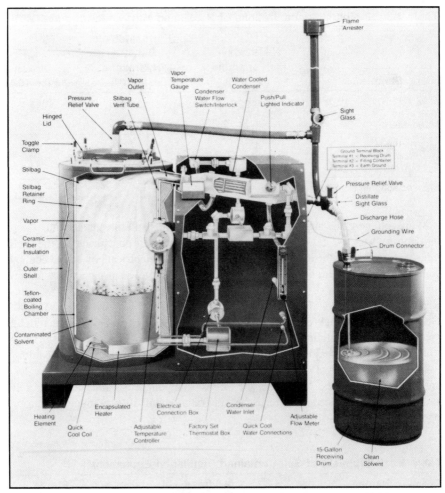

Figure 4.3. A typical atmospheric batch operated still (Photo courtesy of Finish Engineering, Erie, Pennsylvania).

For recycling to be effective, solvents should be segregated. If two or more solvents are mixed, an off-the-shelf still will often be unable to separate them, and a much more expensive, customized unit will be required. Inability to enforce solvent segregation is often the major obstacle to solvent recycling.

Example. Annually, about 25,000 gallons of heptane are used to calibrate aircraft fluid flow sensors at an aircraft maintenance facility. Heptane is used because it is more consistent in performance than jet fuel. Used heptane is stored in an underground waste storage tank.

A nonfractionating-batch atmospheric still was given to this facility with instructions to find a use for it. This responsibility was assigned to a facilities engineer in addition to his regular duties. Local personnel were neither involved in the decision to recycle solvents nor in the selection of the type of still to be used. The engineer decided to use the still to reclaim heptane from the underground waste storage tank. In the still's first test, the recovered solvent failed to meet specifications for use as a calibrating fluid, because of lower (than heptane) boiling and flash points. The contamination was traced to the disposal of solvents other than heptane in the waste storage tank. Segregation of the heptane from other solvents could have alleviated the contamination problems. But, because the engineer was less than enthusiastic about the extra duty and because he believed he had proved that the "still did not work," he made no further efforts. The still was abandoned in place.

Alternatives to Onsite Distillation

If purchasing a still is unfeasible, other methods are available to distill solvents. One method is to contract with a recycler to distill and return spent solvents. Another is to sell the solvents to the recycler. The best method to use depends on the availability of a local recycler, the type of solvent recycled, and the economics of onsite vs offsite recycling. Normally, owning a still is preferable because of cost, process control, and convenience advantages, as well as liability with respect to hazardous waste issues.

One more option is to rent solvent cleaning equipment with a service contract to replace and recycle the solvent.

Safety-Kleen Corporation supplies drums of solvent with self-contained sinks (Figure 4.4). A cleaning solvent is supplied to the sink from a reservoir in the drum,

Figure 4.4.   A rental solvent cleaning system (Photo courtesy of Safety-Kleen, Elgin, Illinois).

and the used solvent drains back to the drum. On a periodic basis, the drum of dirty solvent is replaced with a drum of fresh solvent, and the dirty solvent is distilled at a central facility. Based on its success in supplying cleaning sinks, the company has started marketing a similar facility for cleaning spray paint equipment (Figure 4.5).

Figure 4.5. A rental paint spray equipment cleaner (Photo courtesy of Safety-Kleen, Elgin, Illinois).

With some specialized solvents, the manufacturer will take them back for reprocessing at no charge or for a nominal fee. This recycling can be an advantageous way of disposing of used solvents.

MANAGING A SOLVENT RECYCLING PROGRAM

Waste solvents can be either collected and transported to a centralized distillation facility for recovery or recycled at the point of use. Industrial facilities have been successful with both approaches. Regardless of where the distillation occurs, it is critical that waste solvents be properly segregated and stored so that various solvents and impurities are not mixed.

Centralized Programs for Solvent Recycling

The main advantage of operating a large centralized facility is that capital costs can be recovered quickly because of economies of scale. A centralized facility can redistill large quantities of various types of solvents; there are, however, several disadvantages. Because many different types of solvents are recycled, great care must be taken with waste segregation and sample analysis. Another disadvantage is that solvents must be transported to and from the point of use. A centralized facility depends on an individual dedicated to initiating and supervising operation of the system and an enthusiastic staff dedicated solely to collecting, analyzing, recycling, and distributing the solvent.

Case Study 4.2: Centralized Solvent and Recycle Program

Robins AFB in Macon, Georgia, refurbishes airlift, fighter, bomber, utility, and remote control aircraft as well as helicopters and missiles. The base repairs predominantly C-130 and C-141 transport planes and F-15 fighter jets.

In 1982, Robins AFB purchased a $48,000 batch atmospheric-pressure still manufactured by Finish Engineering Corporation. The still is used to reclaim trichloroethane, freon-113, and isopropanol. In 1983, 227 drums of chemicals were distilled for a savings of $81,000. O. H. Carstarphen, Solvent Reclamation Engineer, estimated that in FY 1984 the recycling of those three chemicals saved the base $118,000 in virgin material and in costs for hazardous waste disposal. The cost to reclaim the used chemicals was only $13 per drum, whereas disposal of the chemicals and repurchase of new materials would have cost from $250 to $500 per drum.

The still can operate up to a temperature of 300°F in the pot and can reclaim organic fluids at a rate of up to 55 gal/hr. Freon and isopropanol have been processed at a rate of approximately 50 gal/hr, and trichloroethane has been processed at a rate of 35–40 gal/hr. Recovery efficiency for isopropanol and freon-113 is approximately 95%. The recovery efficiency for trichloroethane is only 70% because the used material contains nonvolatile waxes, dirt, and greases that are removed from metal parts during degreasing operations.

The Finish Engineering still has been easy and inexpensive to operate and maintain. Some problems were initially encountered with a feed pump when recycling freon, but these have been solved.

Even though reclaimed freon does not meet specifications, it can be used for initial cleaning. Virgin material is then used for final assembly cleaning operations.

Currently, 584 drums of degreasing solvents are used annually by the Directorate of Maintenance. Because it is the predominant solvent used at Robins, approximately 175 drums of trichloroethane per year are currently being reclaimed for reuse in vapor degreasing tanks located in the plating shop. Laboratory tests of the reclaimed trichloroethane have indicated that the material meets military specifications. The Directorate of Maintenance estimated that between 1982 and 1985, recovery of waste trichloroethane saved approximately $79,000.

In the past, isopropanol, which is used for cleaning electronic parts, was discarded when the solution became contaminated with oils and dirt. Currently, however, isopropanol is reclaimed by the organic fluid recovery system, resulting in a savings of $16,200 in FY 1983 and $18,500 in FY 1984. A 5-$\mu$ filter was installed in the discharge line for removal of fine metal particles that were carried over with the alcohol vapors. The reclaimed alcohol had a purity of 99.8%.

Recycling at Robins has been successful because personnel segregate wastes and keep excessive water and other impurities out of the waste slop cans and drums. Segregation of the waste liquids is necessary to maintain the usefulness of the recovered organic fluids. For example, two common paint thinners, MEK and toluene, could easily be mixed together in the waste slop drums in the painting shop. However, if this were to occur, the mixture could not effectively be separated by single-stage batch distillation because the boiling points of the two thinners are similar.

At Robins AFB, management's commitment to the organic fluid recovery operation has been very strong, as demonstrated by the facilities and personnel dedicated to the operation of the system. The Chemical Control Group, consisting of 10 people, collects waste chemicals at 30 different areas. These covered collection areas have controlled access and are on diked concrete pads. The areas are used to dispense fresh solvents from drums and to collect waste solvents in separate, labeled drums. Site managers are responsible for segregating wastes at the different sites.

The Chemical Control Group is also responsible for performing the following tasks: sampling all drums; redistilling freon, trichloroethane, and isopropanol wastes; and transporting the reclaimed materials back to their source. In addition, analytical chemists are required to perform two sets of analyses for each drum of waste. First, as each drum is received, the contents must be analyzed to confirm the labeling. Then, after each distillation run, the recovered solvent is analyzed to ensure that it meets appropriate specifications.

One additional management tool implemented at Robins AFB that has helped the reclamation program succeed is educating base personnel about hazardous wastes. The Directorate of Maintenance developed a course entitled "Storage, Handling, and Disposal of Industrial Chemicals," which is attended by all personnel who store, handle, use, and/or dispose of industrial chemicals. The scope of this training includes industrial materials terminology, personnel protective equipment, hazard identification systems, emergency procedures, and industrial waste collection and disposal.

Case Study 4.3: Centralized Solvent Recycle Program

At Tyndall Air Force Base, Panama City, Florida, solvents are used in the general maintenance of jet aircraft and motor vehicles. In 1981, the Air Force Engineering and Services Laboratory initiated a research project at Tyndall to determine if Stoddard solvent could be economically recycled on the base. The Air Force estimated that approximately 13,000 gal of Stoddard solvent were being used per year at 19 different shops, making it the most widely used solvent at Tyndall in 1981.

A vacuum still, manufactured by Gardner Machinery of Charlotte, North Carolina, was used. This system had a rated capacity of 200 to 225 gal of solvent per hour and was designed to process Stoddard solvent, naphtha, mineral spirits, and petroleum spirits. The solvent recovery system cost approximately $50,000 to purchase and install. The savings dropped from $3.72 per gal of solvent recovered in 1982 to $1.44 per gal in 1983, primarily as a result of a dramatic drop in the price of fresh Stoddard solvent from $4.51 per gal to $1.92 per gal over the same period.[6] Only 4,500 gal of Stoddard solvent were reclaimed, for a savings of approximately $7,000.

The small savings resulted from the system's being underutilized and from the drop in price of virgin material. Many original users switched to a different cleaning solution. Collecting, transporting, and storing the waste Stoddard solvent being generated in the numerous small shops was difficult. Inadequate involvement and commitment of the operational personnel may also have contributed to the limited success of the collection system since the concept had been developed by an outside group and was implemented as a research project. In addition, management's commitment to the success of the project was not as evident as at Warner Robins AFB.

Of the 19 shops that used Stoddard solvent in 1981, only the tire shop actively collected and stored waste solvent for recycle. This shop used two 300-gal dip tanks that contained Stoddard solvent. The cleaning solution was used to remove carbon, grease, and grit from aircraft wheel bearings. Every 4 months, the spent Stoddard solvent was discharged into ten 55-gal drums. The waste solvent in the drums was then pumped to the still holding tank for recycling.

From 1981 to 1984, the still was operated nine times, or approximately 1 day every 4 months. An average of 506 gal of solvent were recycled at a recovery rate of 97% during each of the nine runs. Samples of the recycled solvent were analyzed and generally failed to meet specifications because of an undetected internal leak in the still and a buildup of iron oxide in the system during periods of nonuse.

Because the recycled solvent did not meet specifications, it could not be accepted by the base supply department for distribution and reuse. Most of the recycled solvent was, however, reused in the tire shop, which did not require solvent that met the specifications. Some of the solvent bypassed the supply department and was sent directly to users who expressed an interest in the free material. Although maintenance personnel at the tire shop were pleased with the quality of the recycled solvent, they noticed that the recycled material took longer to dry than fresh solvent.

Operation of the still was discontinued because of its limited use, failure of recycled product to meet specifications, and resultant poor economic performance. In 1985, the still was given to Warner Robins AFB in Macon, Georgia, to supplement its existing solvent recovery unit.

Case Study 4.4: Distillation to Recover Solvent and DI Water

An electronics manufacturer developed an improved method of cleaning soldered semiconductor substrates. The process included cleaning of the semiconductor materials with N-methyl-2-pyrrolidone (NMP), a flux removal agent, followed by rinsing in deionized (DI) water. This process resulted in the production of a mixture of NMP and DI water. CH2M HILL developed, tested, and prepared a conceptual design for a distillation system to recover NMP and DI water from this mixture. In the cleaning process, the NMP solvent becomes contaminated with flux solids, tin, and lead. In addition, the solvent removes minute quantities of oil and grease from the mechanical equipment. Water is introduced into the NMP bath by splashover from the following DI rinse step. The semiconductor substrates carry NMP (and its contaminants) over into the DI rinse. The characterizations of the contaminated NMP and DI waste streams are shown in Table 4.10.

Table 4.10.   Composition and Production of Waste NMP and DI Water

Parameter	NMP	DI WATER
Production (gal/min)	45	15
NMP concentration	99%	1%
Water	1%	99%
Flux	100 ppm	Trace
Metals	Trace	Trace
Oil	20 ppm	Trace

The goal was to develop a feasible process to purify these waste streams; to recover an NMP solvent contaminated with less than 500 ppm of water, 5 ppm of flux, and trace amounts of oil; and to recover a DI stream with less than 500 ppm of NMP.

Purification processes considered along with their advantages and disadvantages are listed on Tables 4.11 and 4.12. Laboratory testing was performed to investigate

Table 4.11.   Evaluation of Potential NMP Purification Processes

Unit Process	Advantages	Disadvantages
Filtration	• Removes solids • Simpler regeneration	• Potential plugging • No removal of water, solubles, or IPA
Precipitation	• Removes metals and solids	• Chemical additions
Ultrafiltration/ reverse osmosis	• Removes dissolved solids	• Membrane durability • Fractional split of components

(cont.)

Table 4.11.   Continued

Unit Process	Advantages	Disadvantages
Crystallization/ cyrogenic	• Likely to render high purity product	• Minimal existing data base • Delicate process
Molecular sieves	• Removes water	• Only applicable to trace removal • Not resistant to upsets • Difficult regeneration and operation • Media durability unknown • IPA may adsorb
Ion exchange	• Removes organic acids • Potential elimination of high boilers • DI water operators understand system • Unattended regeneration	• Resin durability • Complex regeneration • Significant regeneration wastes • Doubtful metals removal • Generates water
Carbon adsorption		• Not selective removal of NMP
Distillation	• Separates NMP from water, metals, and flux • Flexible system • Resistant to upsets	• Operator training • Long startup time
Evaporation	• Removes water	• Partial impurity split
Gas stripping	• Removes water; gas dryness is limiting • Simpler regeneration	• NMP entrainment
Solvent extraction	• Removes water	• Not a binary system • Solvent selection

Table 4.12.   Evaluation of Potential DI Water Purification Processes

Unit Process	Advantages	Disadvantages
Carbon adsorption	• Removes NMP • NMP is recoverable	• Regeneration hysteresis • Carbon durability
Solvent extraction	• Removes NMP	• Further processing required to recover NMP
Ultrafiltration/ reverse osmosis		• Separation unknown • Membrane durability unknown
Distillation	• Removes water from NMP	• Heat of vaporization of water
Molecular sieves		• Water concentration too high

the technical feasibility of the distillation, ion exchange, filtration, and evaporation alternatives for NMP recovery. Laboratory tests were also performed for distillation and carbon adsorption alternatives for DI water recovery.

In the laboratory tests, distillation was shown to be a technically feasible alternative to accomplish the desired recovery and purity of both DI water and NMP. A vacuum distillation system was recommended (Figure 4.6). In this design, both the NMP and the DI water are recovered as distilled condensate. One advantage of this design is that the presence of impurities (such as iron or suspended solids) is minimized in the recovered products.

The estimated capital and operating costs for the design basis system (42 gal/min of NMP and 15 gal/min of DI water) were estimated to be $4,700,000 and $1,700,000, respectively (1981 dollars). Detailed breakdowns of these estimates are provided as Tables 4.13 and 4.14, based on the cost basis provided in Table 4.15.

Localized Solvent Recycling

Localized facilities are sometimes preferable because the waste generator has total control over the recycling operation. Since only a few types of solvents are redistilled at the small facilities, laboratory analysis of waste solvents is often not required. Also, labor intensive transportation and segregation are eliminated. However, decentralized facilities require training more personnel than a centralized facility, as well as convincing them to adopt solvent recovery as part as their routine.

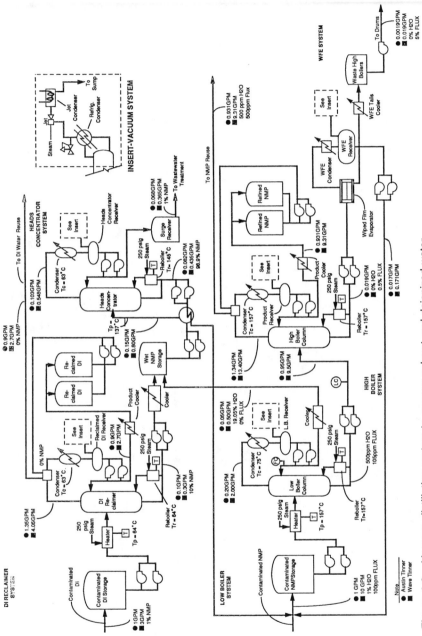

Figure 4.6. Vacuum distillation system to recover solvent and deionized water.

Table 4.13.   Capital Cost Estimate for Vacuum Distillation

Cost Category/[a] System	Storage Tanks and Pumps	DI Reclaimer	Heads Concentrator	Low Boiler	High Boiler	WFE	Total
Equipment	$ 93,800	$ 456,200	$196,600	$245,100	$ 455,000	$160,000	$1,606,700
Installation	18,800	463,700	61,100	144,600	283,100	49,700	1,021,000
Structural	10,300	50,200	21,600	27,000	50,100	17,600	176,800
Mechanical	44,100	214,400	92,400	115,200	213,900	75,200	755,200
Instrumentation	6,600	31,900	13,800	17,200	31,900	11,200	112,600
Electrical	3,800	18,200	7,900	9,800	18,200	6,400	64,300
Subtotal	$177,400	$1,234,600	$393,400	$568,700	$1,052,200	$320,100	$3,746,400
25% Contingency	44,400	308,700	98,400	142,200	263,100	80,000	936,900
Total	$221,800	$1,543,300	$491,800	$710,900	$1,315,300	$400,100	$4,683,200

[a]Total project cost including site development, utility supply, industrial buildings, taxes, and engineering, legal, and administrative fees will be greater than Process Module Total. No allocation for inflation is included.

Table 4.14.   Annual Operating Cost Estimate for Vacuum Distillation

Cost Category/[a] System	DI Reclaimer	Heads Concentrator	Low Boiler	High Boiler	WFE	Total
Energy (steam)	$355,800	$ 13,600	$142,400	$230,700	$ 3,200	$ 745,700
Cooling water	74,800	9,600	12,500	52,600	400	149,900
Maintenance	45,600	19,700	24,500	45,500	16,000	151,300
Waste Disposal losses and solvent losses (1/10%)	—	63,400	140,000	134,200	270,600	608,200
Total	$476,200	$106,300	$319,400	$463,000	$290,200	$1,655,100

[a]Labor costs not included.

Table 4.15. Basis for Operating Cost Estimate

ENERGY COSTS[a]	
Electricity	$0.04/kWh
Steam, 250 psig	$4.00/1,000 lb$_m$
City water, 70°F	$0.18/1,000 gal
MATERIAL COSTS	
Carbon[b]	$1.16/lb$_m$
Ion exchange resin[c]	$205/ft^3
DISPOSAL COSTS[a]	
Waste solvent drumming	$55/55-gal drum
Wastewater treatment[d]	$0.20/lb of BOD
CHEMICAL COSTS[d]	
NMP	$1.15/lb$_m$
Caustic, 10%	$300/ton
Isopropanol	$2.50/gal
DI water	$5.00/1,000 gal
Liquid nitrogen	$0.22/gal

NOTE: Based on 24 hr/day, 320 day/yr, 7,680 hr/yr.
[a]Cost data supplied by client.
[b]Barnebey-Cheney Type PC.
[c]Rohm & Haas Amberlyst A-27.
[d]Cost data supplied by CH2M HILL.

Case Study 4.5: Recovering Painting Solvents with Local Stills

Cleaning operations in the paint shop at Norfolk Naval Shipyard (NSY) generate approximately 15 gal/day of numerous waste solvents including mineral spirits, ketones, and epoxy thinners containing paint pigments. Historically, Norfolk NSY disposed of the waste mineral spirits and other waste organic fluids at a reported cost of $7.80/gal.

Norfolk NSY now uses a nonfractionating batch still, Model LS-15V, manufactured by Finish Engineering, Erie, Pennsylvania. This model is designed to recover 15 gal of solvent per shift of operation (i.e., one full charge of the still pot). The system employs an electrically heated pot with a residue collection pan, a water-cooled shell and tube condenser, a reclaimed solvent collection tank, and an electric vacuum pump. The system is designed to recover organic fluids with boiling points in the range of 100°F to 320°F without using the vacuum system. The vacuum system, which produces a vacuum of 25 in. of mercury during operation, is designed to recover organic fluids with atmospheric pressure boiling points up to 500°F.

The system produces a solid residue in the still pot's residue collection pan. The collection pan is then removed, and the residue is emptied into a container for disposal. The cost of this system (uninstalled) was approximately $9,000. The same system without the vacuum system option cost $5,000.

On the first day of system operation with the vacuum accessory, preparation for startup took only 15 min; the system was started by pressing only one button, and then it ran unattended. On the startup day, mineral spirits were distilled under vacuum. Dry paint solids remained in the collection pan after the cycle was completed and were easily removed for disposal. The system recovered approximately 13 gal of solvent from 15 gal of waste solvent, for an 85% recovery.

The system had also been used successfully without the vacuum system to recover organic fluids with boiling points below 320°F. Norfolk reported recovering more than 50% of the waste solvent at a cost of about $0.05/gal operating at atmospheric pressure.

This solvent recovery operation had three key elements that combined to make it a success: personal dedication of a production representative, technical innovation and ease of operation, and physical location near the waste generation site. Jake Coulter, the paint shop foreman, has been the Champion of this solvent recovery operation. He wanted it to work, and it appears to have been a great success.

Case Study 4.6: Vapor Degreasers with Integral Stills

Anniston Army Depot reconditions used tanks and other armored vehicles. Reconditioning consists of completely disassembling the tanks and dismantling their components. Paint, rust, and dirt are removed from these components prior to remanufacturing. Paint is removed by sand blasting or is stripped by using organic solvents or alkaline strippers. Greases and oils are removed using solvent vapor degreasers followed by alkaline cleaners.

Approximately 15 to 20 trichloroethylene (TCE) vapor degreasers are being used at Anniston Army Depot. All are equipped with a system for distillation solvent recovery. The stills recover TCE from the solvent-oil mixture for reuse in the degreasers. Most stills at Anniston are manufactured by Detrex Corporation. They run continuously when the vapor degreasers are in operation, normally 8 hr/day, 5 days/wk. Dirty solvent is fed from a degreaser boiling sump through a water separator to the recovery still. The steam-heated stills have the capacity to recycle 20 gal/hr of TCE.

Anniston Army Depot has reported no problems in operating and maintaining the distillation units. Twice a year (during shutdown) the vapor degreasers and stills are taken out of service for cleaning and general maintenance. Vapor degreaser TCE baths have never had to be dumped during normal operation or shutdown. TCE is lost through drag-out, evaporation, and disposal in waste still bottoms.

Still bottoms, typically containing 11–17% TCE, oils, greases, and dirt, are automatically discharged to waste holding drums. This hazardous waste is sent to a commercial contractor for treatment. Anniston investigated the cost-effectiveness of recovering TCE from still bottoms and determined that the still bottoms would have to contain 40% TCE before it would be economical to recover additional solvent.

REFERENCES

1. Cheng, S. C., et al. "Alternative Treatment of Organic Solvents and Sludges From Metal Finishing Operations," U. S. Environmental Protection Agency, EPA-600/2-83-094, September 1983.
2. Johnson, J. C., et al. "Metal Cleaning by Vapor Degreasing," *Metal Finishing*, September 1983.
3. Kohl, J., P. Moses, and B. Triplett. "Managing and Recycling Solvents," Industrial Extension Service, School of Engineering, North Carolina State University, Raleigh, NC, December 1984.
4. "Vacuum Still Operation Manual." Gardner Machinery Corporation. Charlotte, NC.
5. Joshi, S. B., et al. "Methods for Monitoring Solvent Conditions and Maximizing Its Utilization," paper presented at the 8th ASTM Symposium on Hazardous and Industrial Solid Waste Testing and Disposal, Clearwater, FL, November 12–13, 1987.
6. Tapio, G. E. "A Limited Test of Solvent Reclamation at an Air Force Base," AFESC/RDV, Tyndall Air Force Base, Panama City, FL, 1984.

GENERAL REFERENCE

Isooka, Y., Y. Imamura, and Y. Sakamoto. "Recovery and Reuse of Organic Solvent Solutions," *Metal Finishing*, June 1984.

CHAPTER 5

Metal Plating and Surface Finishing

DESCRIPTIONS OF METAL FINISHING OPERATIONS AND WASTES

Metal finishing operations involve preparing and finishing metal parts for final use. Metal finishing operations can include cleaning, degreasing, pickling (acidic removal of surface oxides), electroplating and electroless metal plating, etching, and conversion coating. These processes involve applying a functional, protective, or decorative coating to a metal part (or in circuit board manufacture, adding a metallic coating to a plastic substrate) to add value to that product. Cleaning processes are used to prepare parts for coating and to improve the adhesion of that coating to the surface.

Metal finishing operations produce waste streams containing acids and bases, toxic heavy metals, and solvents and oils. Table 5.1 shows a list of waste components typically found in metal finishing operations.

Metals and solvents are the principal components regulated under wastewater treatment and hazardous waste regulations.

Table 5.1. Metal Finishing Operations and Typical Wastes

Source	Waste
Degreasing	Solvents, oils
Cleaning	Alkalis, metals, chelates, solvents
Pickling	Acids, metals, chromates
Metal plating	Acids, metals, cyanide, alkalis, chelates
Etching	Metals, acids, chelates
Conversion coating	Chromate, phosphates, metals

Chrome Plating

Chrome plating is performed for one of three reasons: to enhance the corrosion resistance; to provide a bright metallic appearance; or to impart improved mechanical properties (hardness, lubricity) to the underlying base metal. Plating for corrosion protection or appearance is usually referred to as "decorative chrome plating" and generally involves adding a thin (a few thousandths of an inch) coating to a part. This process can be accomplished in a matter of minutes. Many parts are plated in a single tank, often using automated equipment, with large volumes of plating solution being "dragged out" of the plating bath each day.

Plating to improve mechanical properties is usually referred to as "hard-chrome plating." Hard-chrome plating often involves adding tens of thousandths of an inch thickness of plate, a process that can take more than 24 hr to complete. As a result, hard-chrome plating drag-out rates are significantly less than the rates for decorative plating.

Nickel, Cadmium, and Zinc Plating

Nickel, cadmium, and zinc are also used for plating parts to provide a corrosion protection finish. These coatings are significantly thinner than hard-chrome plates and are applied in minutes, rather than the hours or days required for hard-chrome plating. Nickel is applied to new parts for corrosion and wear resistance, as well as to rebuild worn parts. A thin nickel plate is sometimes applied prior to hard-chrome plating.

Sacrificial cadmium and zinc coatings are normally applied to protect the base metal, typically iron or steel. A thin surface coating is usually applied for corrosion protection, for improvement of wear or erosion resistance, for reduction of friction, or for decorative purposes. Since cadmium is significantly more expensive and toxic than zinc, it is used as a protective electroplate only in those circumstances requiring its special properties.

Cadmium is often selected over zinc as a protective coating for the following reasons: (1) it is more easily soldered than zinc; (2) its corrosion products do not swell and are not bulky (unlike the "white rust" formed by zinc) and hence do not interfere with functional moving parts; (3) cadmium plating is easier to control than zinc plating; and (4) cadmium is somewhat superior to zinc in corrosion protection in marine (salt) environments. Parts that are to be cadmium-plated typically are cleaned of grease, dust, oil, and rust by undergoing solvent vapor degreasing, alkaline cleaning, and acid pickling. After a part is cleaned, it is cadmium-plated and then heated to remove hydrogen (to prevent hydrogen embrittlement).

Copper, Gold, and Silver Plating

Because of copper's electrical conductive properties, copper is plated on plastic in the manufacture of printed circuit boards. Gold and silver are plated on electrical contacts and high-value circuits because of their superior physical properties, such as higher conductance and inertness. Gold and silver are also plated on products for aesthetic appeal.

Printed Circuit (Wiring) Board Manufacture

Production of printed circuit boards involves the plating and selective etching of flat circuits of copper supported on a nonconductive sheet of plastic. Production begins with a sheet of plastic laminated with a thin layer of copper foil. Holes are drilled through the board using an automated drilling machine. The holes are used to mount electronic components on the board and to provide a conductive circuit from one layer of the board to another.

Following drilling, the board is scrubbed to remove fine copper particles left by the drill. The rinsewater from a scrubber unit can be a significant source of copper waste. In the scrubber, the copper is in a particulate form and can be removed by filtration or centrifuge. Equipment is available to remove this copper particulate, allowing recycle of the rinsewater to the scrubber. However, once mixed with other waste streams, the copper can dissolve and contribute to the dissolved copper load on the treatment plant.

After being scrubbed, the board is cleaned and etched to promote good adhesion and then is plated with an additional layer of copper. Since the holes are not conductive, electroless copper plating is employed to provide a thin continuous conductive layer over the surface of the board and through the holes. Electroless copper plating involves using chelating agents to keep the copper in solution at an alkaline pH. Plating depletes the metal and alkalinity of the electroless bath. Copper sulfate and caustic are added (usually automatically) as solutions, resulting in a "growth" in volume of the plating solution. This growth is a significant source of copper-bearing wastewater in the circuit board industry.

Treatment of this stream (and the rinsewater from electroless plating) is complicated by the presence of chelating agents, making simple hydroxide precipitation ineffective. Iron salts can be added to break the chelate, but only at the cost of producing a significant volume of sludge. Ion exchange is used to strip the copper from the chelating agent, typically by using a chelating ion exchange resin. Regeneration of the ion exchange resin with sulfuric acid produces a concentrated copper sulfate solution without the chelate. This regenerant can then be either treated by hydroxide precipitation, producing a hazardous waste sludge, or else concentrated to produce a useful product.

Growth from electroless copper plating is typically too concentrated in copper to treat directly by ion exchange. Different methods have been employed to reduce the concentration of copper sufficiently either to discharge the effluent directly to sewer or to treat with ion exchange. One method is reported by Hewlett-Packard[1] in which growth is replenished with formaldehyde and caustic soda to enhance its autocatalytic plating tendency, and then mixed with carbon granules on which the copper plates out in a form suitable for reclaiming.

Following electroless plating, copper is electroplated on the board to its final thickness, and a layer of tin-lead solder is plated over the copper. A photoresist material is then applied to the board and exposed by photoimaging a circuit design. Following developing and stripping a selected portion of the photoresist, that portion of the tin-lead plate is etched to reveal the copper in areas other than the final desired circuit pattern.

The exposed copper is then removed by etching to reveal the circuit pattern in the remaining copper. Ammonia-based etching solutions are most widely used. Use of ammonia complicates waste treatment and makes recovery of copper difficult. An alternative to ammonia etching is sulfuric acid/hydrogen peroxide etching solutions. This latter etchant is continuously replenished by adding concentrated peroxide and acid as the copper concentration increases to about 80 g/L. At this concentration, the solution is cooled to precipitate out copper sulfate. After replenishing with peroxide and acid, the etchant is reused. Disadvantages of the sulfuric acid-peroxide etching solution are that it is relatively slow when compared with ammonia, and controlling temperature can be difficult.

Preparation for Metal Finishing

Preparation for metal plating produces waste streams from several sources. Precleaning and removing surface oxides (pickling) in process baths result in metal-bearing acidic and caustic waste streams. Each operation is typically followed by a rinsing operation that produces a dilute waste stream. Spent or contaminated process baths constitute a concentrated, intermittent waste source.

The metal finishing process itself produces several waste streams (Figure 5.1). Following each step in the process, parts are rinsed to remove finishing solutions that adhered to the parts (drag-out). Most plating operations use single-overflow rinse tanks that operate at flow rates of 2 to 8 gal/min. Rinsewater flows are typically the predominant sources of wastewater at plating facilities. Additional discharges of hazardous waste include cleanup of spills; aerosol spray from operations such as chromium plating, which is exhausted to the atmosphere or removed by wet scrubbers; and discarded process solutions.

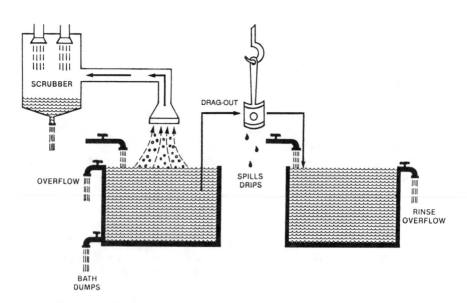

Figure 5.1. Sources of waste from metal finishing processes.

In metal finishing operations, wastewaters are typically produced from many separate plating or finishing processes. The result can be production of a mixed wastewater that contains several metals and chelating agents. Conventional metal hydroxide precipitation is unworkable for treating these metal-containing wastewaters.

TECHNIQUES FOR REDUCING HAZARDOUS WASTE GENERATION

Several process modifications have been proposed to reduce, at their source, hazardous waste generation from metal plating. These modifications include:

- improving housekeeping practices
- reducing drag-out
- modifying rinsing
- recovering metals from rinsewaters
- reducing or eliminating tank dumping
- substituting less hazardous materials

Each modification is discussed in detail in this section.

Improving Housekeeping Practices

Changes in housekeeping practices can be made rapidly with little capital investment. Successfully implemented, these changes can result in increased production rates, improved product quality, and improved workplace safety, as well as reduction of hazardous waste generation. Reducing hazardous waste generation can yield significant savings in raw material usage and wastewater treatment. The following list of housekeeping practices, although not all-inclusive, could save plating shops thousands of dollars a year:

- Repair all leaking tanks, pumps, valves, etc.
- Inspect tanks and tank liners periodically to avoid failures that may result in bath dumps. Inspect steam coils and heat exchangers to prevent either accidental contamination or leakage of steam condensate and cooling water into the plating bath.
- Install high-level alarms on all plating and rinse tanks to avoid accidental bath dumps.
- Maintain plating racks and anodes to prevent contamination of baths. Remove racks and anodes when baths are not in use.
- Minimize the volume of water used during cleanup operations.

- Properly train plating personnel so that they understand the importance of minimizing bath contamination and wastewater discharge.

- Thoroughly clean and rinse parts prior to plating to minimize contamination of the plating bath. Areas that are not to be plated should be masked or stopped off with tape or wax to limit corrosion from these areas. Parts should be removed from the bath when not being plated.

- Remove anodes from tanks when plating is not being performed. In baths where erosion of the anodes provides replacement metal, dissolution of anodes during periods of nonuse can result in a buildup of metal to a concentration higher than acceptable. This buildup often results in the need to dispose of a portion of a bath to reduce the metal concentration.

Reducing Drag-Out

Reducing drag-out (formation of film on part surface) from plating baths saves bath replacement chemical costs and reduces waste disposal costs. To evaluate the effectiveness of drag-out reduction, existing drag-out must be quantified. For example, the drag-out from barrel plating tanks is usually 10 times greater than that removed from baths employing rack plating. The shape and design of the parts, racks, and barrels can also significantly affect drag-out rates. A cup-like depression in a part or rack can literally bail gallons of plating solution from a bath. A more favorable rate of return is realized by implementing drag-out reduction techniques at plating lines for decorative chrome, cadmium, and zinc. In those lines, plating times are relatively short and drag-out is significantly greater than in hard-chrome plating.

Drag-out can be reduced by decreasing either bath viscosity or surface tension. Viscosity can be lowered by reducing the chemical concentration of the bath or by increasing the temperature of the bath. Surface tension can be reduced by either adding nonionic wetting agents or increasing bath temperature. These modifications will either improve the drainage of plating solutions back into plating baths or reduce the concentration of metal in the drag-out. Lowering the velocity of withdrawal of parts from a bath can drastically reduce the thickness of a drag-out layer because of surface tension effects.

Many production platers use automated plating lines, which consist of a series of tanks in which plating racks are operated by a robotic crane. This type of operation is standard in the circuit board industry. One circuit board manufacturer reprogrammed its automated rack mover to increase the holding time over the plating baths, significantly reducing drag-out without adversely affecting production.

Drag-out can be captured by the use of drain boards, drip bars, and drip tanks and can be returned to the bath (Figure 5.2). These simple devices save chemicals, reduce rinse requirements, and prevent unnecessary floor washing.[2] Significant drag-out reduction can be accomplished if platers carefully rack and remove parts to minimize entrapment of bath materials on surfaces and in cavities (Figure 5.3). Parts should be designed to promote drainage (Figure 5.4).

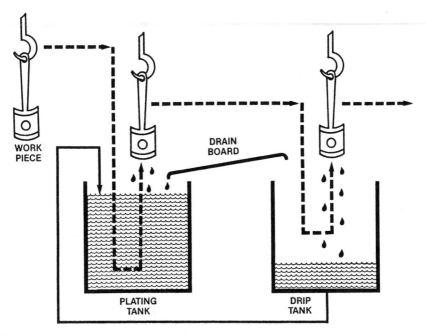

Figure 5.2. Drag-out recovery devices.

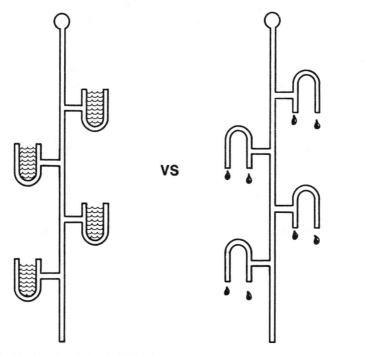

Figure 5.3. Racking to minimize drag-out.

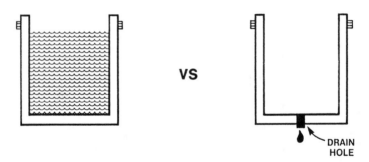

Figure 5.4. Part design to minimize drag-out.

Air knives can be used to knock plating films off parts and back into process tanks. This technique is particularly effective in removing ambient temperature solutions from plated parts.

Modifying Rinsing

Rinsing is used to remove residual drag-out from parts and racks. Rinsewaters must be sufficiently clean to reduce the concentration of these chemicals in a reasonable period of time. Most platers and surface finishers employ continuously flowing single-tank rinses to wash out contaminants. Rinse flows are typically controlled manually and left running continuously. Flows are usually set to eliminate ''color'' in the water, rather than using a chemical analysis.

Decreasing the rinsewater flows may not reduce the amount of toxic metals to be disposed of, but it can reduce the volume of liquid waste that must be processed in industrial wastewater treatment plants. However, concentrations of metals would increase, resulting in possible adverse effects on treatment. Thus, decreasing rinsewater flows may not appreciably reduce costs of wastewater treatment, especially if treatability is impaired.

If the rinse flow rates are reduced sufficiently, it is possible to use rinsewater to make up for evaporative losses in the plating tanks, resulting in metal recovery and reduced waste discharge. Reducing flows can also increase the efficiencies of metal recovery by using concentration processes such as evaporation, ion exchange, and reverse osmosis (discussed in a later section).

The following are descriptions of techniques that improve rinse efficiency.

Rinsing Over Plating Tanks

Rinsing over the plating tanks also effectively removes drag-out from parts. A plated part is held over the plating tank and sprayed with rinsewater. More than 75% of plating chemicals drain back to the plating bath. This modification is best suited for parts that are plated at elevated temperatures so that evaporation rates are high, thus making space available for the amount of rinsewater poured into the tank.[3]

Example. At NASA's Marshall Space Flight Center, the plater simply rinsed plated parts over the plating bath with a handheld hose, recovering the chemical and replacing water lost to evaporation in the bath. This simple but effective technique reduced the concentration of plating metals in the rinsewater to less than 0.1 mg/L.

Spray Rinse

A spray or fog rinse can be used to improve the efficiency of rinsewater use. Drainage can be directed back into the process tank, if evaporation is sufficient, or into a drag-out tank. This modification is best suited for flat parts, such as circuit boards, since getting into recesses is difficult without submersion and good mixing.

Still Rinse

Still or "dead" rinse tanks (Figure 5.5) can be used prior to the use of rinse tanks with flowing clean water. Water from the drag-out tank or still rinse tank can be returned to the bath to make up for evaporation losses. Increasing plating bath temperatures to increase evaporation may be justified.

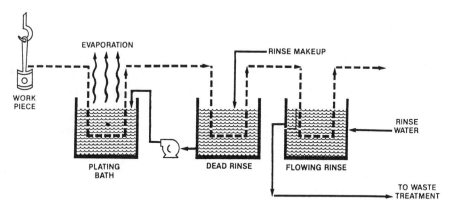

Figure 5.5. Dead rinse for recovery of plating chemicals.

Example. At a small plating shop at Williams Air Force Base several years ago, the plater kept a plastic bucket of distilled water next to each plating bath. After plating, he would dip each rack of parts in the bucket before final rinsing in a flowing rinse tank. Then, before starting work each morning, he would empty the bucket from the previous day into the plating tank, recovering the chemical, and refill his bucket with distilled water.

Flow Control

Water supply control valves can be used to reduce wastewater flows to a minimum. These inexpensive devices (approximately $30) regulate the feed rate of fresh water within a narrow variation of flow despite variations in line pressure. These controllers can usually be set to regulate flow within a 0.5-gal/min range. One difficulty with limiting flow to a rinse tank is that water agitation is essential to good rinsing, and a high flow of water often provides agitation to the rinse tank.

Conductivity Control

Conductivity controllers (Figure 5.6) can be used to operate rinsewater control valves, thereby reducing demands on plating personnel. Conductivity control operates on the principle that clean water has a lower conductivity than water contaminated with plating solutions. A conductivity probe, controller, and valve reportedly can cost less than $1,000 to purchase and install;[4] systems have been installed in many plating shops. Unfortunately, these units have not performed well in most installations because of the probes' lack of ruggedness and need for frequent calibration and cleaning. Selecting the optimum conductivity setpoints can be difficult. Platers frequently override or disconnect conductivity controllers because of dissatisfaction with their operation.

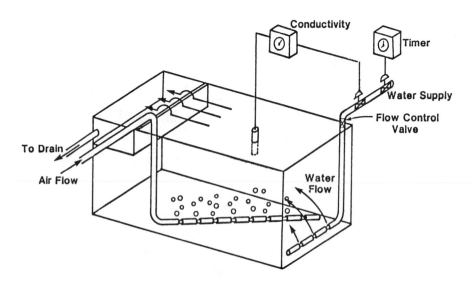

Figure 5.6. Use of controls and air agitation to reduce rinsewater usage.

Example. An aerospace company manufactures experimental circuit boards at its New England facility. Because of the experimental nature of the operation, only a small number of circuit boards are manufactured on a full-size plating line. When conductivity controllers were installed on all of the rinse tanks, the platers discovered that the controllers could not be set accurately enough to turn on the rinse flow each time a rack was lowered into the rinse tank, which was unacceptable since they relied on fresh water flow to provide mixing. To remedy the problem, conductivity controllers are being replaced by micro-switches on the rack holders. These switches will activate valves to provide rinsing whenever a rack is lowered into the tank. In addition, aeration is being added to improve mixing, thus increasing the efficiency of rinsing even at reduced water flows.

Improved Rinse Tank Mixing

Improvements to rinse tank mixing can increase the efficiency of water use (Figure 5.6) and, therefore, allow a reduced flow of water that normally would not be acceptable. A submerged influent water line evenly distributes fresh water through the tank and creates a rolling action. Most improvements in rinse tank mixing are effected by aeration. Existing facilities can be retrofitted with these modifications using low-pressure blowers and inexpensive plastic (PVC) piping.

Timers

Timer controls can be added that will reduce the daily volume of rinsewater while at the same time providing an adequate flow to maintain rinsewater quality and assure adequate mixing when rinsing is required. The plater no longer has to remember to shut off the valve after rinsing (Figure 5.6). For reasonably uniform plating operations, timers can be used to operate rinse valves on a preset cycle. For intermittent plating operations, timers can be installed that can be initiated manually, providing an adequate flow when needed for mixing and an adequate period of flow to reduce the concentration of contaminants for the next batch. For automated systems, rinsing can be controlled by the automatic sequencer.

Example. For a California company, CH2M HILL recommended that manually activated timers be installed on the rinse tanks. Pushing a convenient button at the front of each tank will provide up to 1 hr of flow to the rinse tank. Equipment costs for switch, timer, valve, and controls total approximately $400 per tank.

Cascade Rinse

Cascade rinsewater recycling is the technique in which overflow from one rinse tank is used as the water supply for another compatible rinsing operation. For example, rinsewater effluent from an acid dip tank can be cascaded to an alkaline cleaner rinse tank. Interconnecting rinsing tanks can complicate operations, but the savings often exceed the additional operation cost.

Countercurrent Rinsing

In countercurrent rinsing, parts are sequentially immersed in a series of tanks, countercurrent to the rinse flow (Figure 5.7). Countercurrent multiple rinse tanks can reduce rinse flows by over 95% compared to single overflow rinses. Optimum countercurrent rinsing usually employs three tanks operating in series.

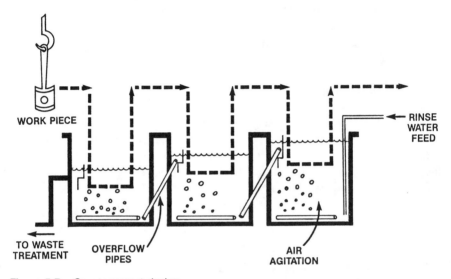

Figure 5.7.   Countercurrent rinsing.

The concentration of plating solution in each successive rinse tank can decrease by a factor of 10. For example, assume that the drag-out concentration of a plating bath contains 40,000 mg/L of dissolved solids, and the final rinse is limited to 40 mg/L. Concentrations of dissolved solids in the three multiple rinse tanks could be controlled to 4,000, 400, and 40 mg/L. For a drag-out rate from the plating bath of 1 gal/hr, a countercurrent rinse flow of 10 gal/hr would be sufficient, as compared to 1,000 gal/hr for a single rinse tank.

Many plating facilities do not include countercurrent rinsing because the required additional space is not available. Another consideration is the additional production time because parts must be rinsed in more than one tank. Where space is available, the cost of additional rinse tanks can range from $1,000 to $10,000 per tank, depending on size, shape, and construction materials.

It is sometimes difficult to provide adequate mixing in countercurrent tanks because, when compared with conventional single rinse tanks, flows are considerably reduced. Aeration is necessary to provide sufficient mixing.

Example. At one facility, countercurrent rinsing was considerably less efficient than is normally expected. Aeration to provide mixing caused a backflow of rinsewater to adjacent rinse tanks, resulting in considerably lower water savings than predicted. When the tanks could not be balanced to provide adequate mixing without backflow, countercurrent rinsing was abandoned.

Countercurrent rinse systems can be retrofitted in existing tanks by adding baffles, weirs, pipes, and pumps. Savings vary considerably as a result of differences in costs of raw water and wastewater treatment. At many facilities, the payback period can be as short as 1 year. When plating solution is recovered, further savings can be realized by returning the most concentrated rinsewater to the plating bath to make up for evaporative losses. Similar savings can be accomplished by employing a dead or still rinse, followed by a flowing rinse. The contents of the still rinse are periodically returned to the plating bath to recover the plating chemicals.

Recovering Metals From Rinsewaters

Evaporation, ion exchange (IX), reverse osmosis (RO), and electrodialysis (ED) have been used to recover chemicals from rinsewaters. These processes reconcentrate plating solutions from rinsewater, producing a relatively pure water, which is reused for rinsing. Both general and site-specific factors must be evaluated to determine the recovery process best suited for a particular plating operation. Factors include the type of metal being plated, drag-out rates, rinsewater concentrations and flows, space requirements, staffing requirements, availability of utilities (such as steam or electricity), and costs for water and wastewater treatment and for sludge disposal.

Evaporation

Evaporation is the oldest method used to recover plating chemicals from rinse streams. In this process, enough rinsewater is boiled off to concentrate the solution sufficiently to return it to the plating bath. The steam can be condensed and reused for rinsing. Evaporators can be operated under a vacuum to lower the boiling temperature, thus reducing energy consumption and preventing thermal degradation of plating additives.

The degree of concentration required of the evaporator can be reduced by increasing the evaporation rate from plating baths. Raising the operating temperature can significantly increase the evaporation rate but only at the expense of added heating costs. Use of air agitation in a plating tank can also increase the surface evaporation rate.

Because of their high energy use, evaporators are most cost-effective in concentrating rinsewaters that are to be returned to hot baths, such as those used in chromium plating, where high evaporation rates reduce the concentration required.

Evaporative recovery also has been used successfully for ambient temperature nickel baths and various metal cyanide baths. The capital and operating costs of an evaporator can be reduced by using countercurrent rinsing to produce a low-volume, concentrated rinse stream. One study estimated that chrome-plating shops at Naval shipyards could save $17,000 a year (1983 dollars) by using a countercurrent rinse system in conjunction with evaporative recovery.[5] The payback period was estimated to be less than 1 year (Figure 5.8). A typical atmospheric evaporator is shown as Figure 5.9. Table 5.2 lists suppliers of atmospheric evaporators.

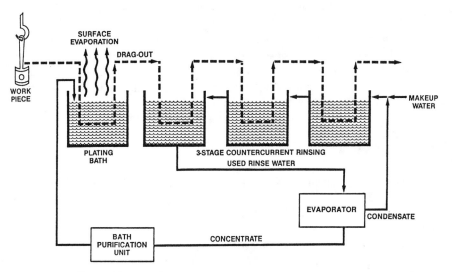

Figure 5.8.   Evaporative metal recovery.

Table 5.2.   Suppliers of Atmospheric Evaporators

Company	Address	Phone
Brucar Process	127M Brook Avenue Deer Park, NY 11729	(516) 586-5800
ERC/Lancy	525 West New Castle Avenue Zelienople, PA 16063	(412) 452-9360
Harshaw/Filtrol	3915 D Valley Court Winston-Salem, NC 27106	(800) 321-4802
Poly Products	P.O. Box 151 Atwood, CA 92601	(714) 538-0701
Techmatic, Inc.	133 Lyle Lane Nashville, TN 37210	(615) 256-1416
Water Management	2300 Highway 70 East Hot Springs, AR 71901	(501) 623-2221

Ion Exchange

Ion exchange (IX) uses charged sites on a solid matrix (resin) to selectively remove either positively charged ions (cations) or negatively charged ions (anions)

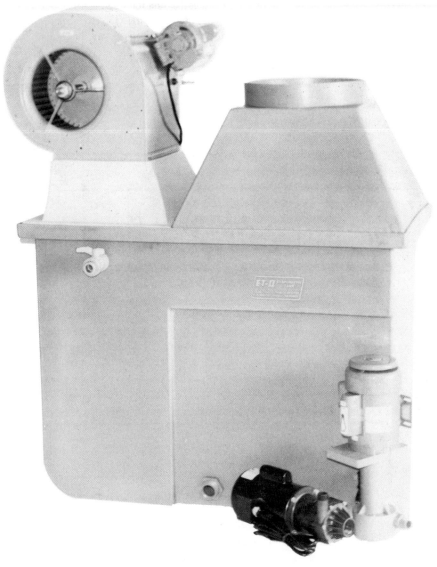

Figure 5.9. Atmospheric evaporator (Photo courtesy of Poly Products Corporation, Atwood, California).

from the solution. Ions removed from the solution are replaced by an equivalent charge of ions displaced from the resin, hence the name ion exchange. Exchanged rinsewater is normally recycled.

Following saturation of the exchange sites, ion exchange resins are usually regenerated by passing an acid or base through them, producing a concentrated metal solution that can be recycled.

In metal plating operations, anionic exchange resins have been used to recover chromic acid from rinsewaters, typically exchanging hydroxide ions for the negatively charged chromic acid anions (Figure 5.10). Anionic resins have also been used to recover cyanide and metal cyanide complexes. Cationic exchange resins have been used to recover metal cations. An IX system typically consists of a wastewater storage tank, prefilters, cation or anion exchanger vessels, and caustic or acid regeneration equipment.

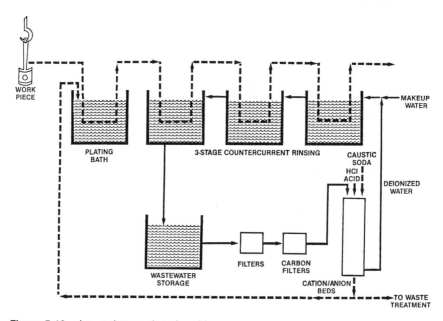

Figure 5.10. Ion exchange chromic acid recovery.

In general, ion exchange systems are suitable for chemical recovery applications where the rinsewater has a relatively dilute concentration of plating chemicals and where a relatively low degree of concentration is required for recycle of the concentrate. The recovery of plating chemicals from acid-copper, acid-zinc, nickel, tin, cobalt, and chromium plating baths has been commercially demonstrated. The process has also been used to recover spent acid-cleaning solutions and to purify plating solutions for longer service life.

An EPA study estimated that an IX system being operated 5,000 hr/yr would cost $31,000 to install and $6,000/yr to operate (in 1979 dollars), resulting in a 5.2-yr payback period.[6] Ion exchange recovery systems are not cost-effective, however, when drag-out rates are low. According to an EPA study, a favorable payback period of 2.8 yr was estimated for chromic acid recovery from rinsewater where the chromic acid drag-out rate is 3 lb/hr.[7] For drag-out rates significantly lower (e.g., those used in hard-chrome plating) an IX recovery system is not normally cost-effective.

IX may also be uneconomical where wastewater treatment and sludge disposal costs are minimal.

A reciprocating flow ion exchanger (RFIX) is the most widely used IX system for the recovery of chemicals from plating rinses. These proprietary skid-mounted units are specially designed to purify plating rinsewaters. The units cost less and require less space than conventional fixed-bed systems, and they incorporate regenerative chemical reuse techniques to reduce operating costs and to yield higher product concentration for recycle. RFIX units have proved effective in three basic applications:

- recovery of chromic acid from rinsewaters

- recovery of nickel, copper, zinc, tin, and cobalt from rinsewaters

- concentration of mixed-metal rinse solution for disposal

IX has been most successful when recovering chromic acid and nickel from rinsewaters, but problems have been encountered in concentrating mixed-metal solutions. By using the ion exchanged water for rinsing, fresh water consumption can be reduced by 90%. However, waste regenerant brine can be difficult and expensive to treat and dispose of. Often, the environmental and economic benefits of reduced water consumption can be offset by an increased use of treatment chemicals.[8] Table 5.3 lists suppliers of ion exchange equipment.

Table 5.3. Suppliers of Ion Exchange Equipment

Company	Address	Phone
Crane	800 3rd Avenue King of Prussia, PA 19406	(215) 265-5050
Eco-Tec	925 Brock Road, South Toronto, Ontario L1W 2X9	(416) 831-3400
Graver	2720 US Highway 22 Union, NJ 07083	(201) 964-0768
IWT	4669 Shepherd Trail Rockford, IL 61105	(815) 877-3041
Met-Pro	163 Cassell Road Harleysville, PA 19438	(215) 723-6751
Penfield	8 West Street Plantsville, CT 06479	(203) 621-9141
Permutit	East 49 Midland Avenue Paramus, NJ 07652	(201) 967-6000
Sethco	70 Arkay Drive Hauppauge, NY 11788	(516) 435-0530

The following case study illustrates the combination of IX and evaporation for the recovery of chrome from hard-chrome plating.

Case Study 5.1: Vapor Recompression Evaporation for Chrome Recovery at Charleston Naval Shipyard

Industrial process description. The Naval Shipyard (NSY) at Charleston, South Carolina, employs approximately 8,000 people to repair, refurbish, and recondition naval surface ships and fossil- and nuclear-fueled submarines.

The plating shop at Charleston NSY performs several operations, including hard and flash chromium; cadmium, copper, nickel, and zinc electroplating; silver brush plating; plasma spray and hot dip galvanizing; stripping of chromium, nickel, and copper; application of chromate conversion coatings; phosphating; electropolishing; and passivation. Items hard-chrome plated at Charleston consist primarily of functional parts such as rotors, hydraulic cylinders, bearing caps, shafts, and end bells from mechanical and electrical machinery on ships.

Parts are hard-chrome plated to an average plated thickness of approximately 20 thousandths of an inch, with a range of 1 to 40 thousandths. All parts are overplated and subsequently ground to final dimensions. To achieve the thickness required, parts are plated for an average of 27 hr.

At Charleston, hard-chrome plating is performed in 13 plating tanks with a 125-ft^2 combined surface area. These plating baths are significantly underutilized. Long plating times and job shop conditions have caused low production rates. Usually fewer than 150 parts have been hard-chrome plated each month (12 parts per tank). Low production rates and overcapacity have resulted in extremely low drag-out rates (0.08 gal of plating bath/hr). Typical bath composition and operating conditions used at the Charleston plating shop are shown in Table 5.4.

Table 5.4.   Charleston NSY Hard-Chromium Plating Operating Conditions

Parameter	Value
Chromic acid concentration	32–40 oz/gal
Chromic acid/sulfate ratio	100:1–125:1
Bath temperature	100–140°F
Aeration rate	8–15 ft^3/min

Note: Anode is insoluble lead (lead-tin, lead-antimony).

Process modification description. A study was initiated to evaluate a method to reduce or eliminate chromium waste discharges from the plating shop by recovery of chromium from rinsewaters. The system tested, an ion exchange/evaporator manufactured by LICON, Inc., was designed to recover chromium from up to 33 gal/hr of hard-chrome plating rinsewater. The LICON system consisted of two basic modules: a cleanup ion exchange module used to remove cations (principally iron and trivalent chromium) from dilute rinsewater and an evaporation unit to concentrate the cation-free rinsewater to plating bath strength. The evaporator unit utilized vapor recompression and waste heat recovery to reduce energy consumption. The final product was to be a clean rinsewater concentrate of plating bath strength and a condensate of distilled water quality as a beneficial by-product. A portion of the cooled distilled water is used to cool and seal the lobes of the vapor compressor.

Process modification experience. The LICON unit was originally tested at the Pensacola Naval Aviation Depot (NADEP). An evaluation of the LICON unit's performance at that facility was prepared by Charles J. Carpenter.[9] Annual savings of 35,000 lb of chromium had been projected for the unit at this installation, based on the assumption that the drag-out rates and rinsewater chromium concentrations would be the same as those of commercial decorative chrome platers.

During the year and a half of operation at Pensacola NADEP, the LICON unit cost approximately $195 to operate each day, produced 552 gal of poor quality distilled water, and recovered approximately a half pound of chromium, which was unsuitable for reuse. The initially forecast recovery rate did not take into account the differences between hard-chrome plating and decorative plating. Because of the long plating times and low production rates, drag-out to rinsewater was orders of magnitude less for Navy hard-chrome plating than for commercial decorative plating. At Pensacola NADEP the maximum chromium drag-out available for recovery was determined to be approximately 90 lb a year. In summary, operation of the LICON unit at Pensacola was uneconomical, costing approximately $1,500 per lb of chromium recovered,[9] compared with a replacement cost of less than $2 per lb for new chrome plating solution. Carpenter also performed a reliability, availability, and maintainability study.[9] The reliability and operational availability of the LICON unit were rated as very poor, principally because of problems with the vapor recompression unit. In summary the report stated:

> The LICON vapor recompression unit appears sound in theory, but needs more work before it will be a reliable, viable alternative to consider for use in metal recovery. . . . Economically, the LICON unit is a liability at NARF [NADEP] Pensacola.

The LICON evaporator with the vapor recompression unit and cation exchange module was relocated from Pensacola to Charleston Naval Shipyard in South Carolina for further testing and evaluation on the hard-chrome plating line at that facility. A new contract for refurbishing and installing this equipment and an additional maintenance agreement were negotiated with LICON. As part of the refurbishment, a new compressor was installed in the evaporator.

The feasibility of using the LICON unit for chromium recovery was evaluated over a 9-month period during which information was collected concerning the costs and level of effort for installation, startup, operation, and maintenance. In addition to recovery of chromium from rinsewater, the LICON unit was also used to clean up contaminated plating baths. The cation exchange module was tested by itself to remove cations from rinsewater for plating bath makeup. In addition, the vapor recompression evaporator was used to concentrate mixed plating wastes and thus reduce their disposal costs.

Baker S. Mordecai[10] evaluated the LICON unit's performance at Charleston. The report evaluated the feasibility of using the LICON unit for three separate tasks:

1. Evaporative Recovery—use of the vapor recompression evaporator and cation exchange unit for hexavalent chromium recovery

2. Rinsewater Cleanup Using Cation Exchange—use of the cation exchange unit for removing cationic contaminants from the rinsewater, which was then used for evaporation makeup and spray rinsing into the hard-chrome plating tanks

3. Mixed Rinse Concentration by Evaporation for Disposal—use of the evaporator for concentrating the metals from several other plating bath rinses as an alternative to conventional treatment for disposal

The results of the evaluation of the LICON unit and associated parts are discussed below, along with a summary of the unit's role in cleaning up concentrated plating solutions. It should be noted that, although a recycle and reuse system was developed that essentially eliminated the need to concentrate the chrome plating rinsewater, personnel at Charleston continued to evaluate this function of the LICON unit in the interest of giving it a fair trial.

Evaporative recovery. Major problems were encountered during operation of the LICON vapor recompression concentrator. First, the ductile iron compressor became severely corroded because of the acidity of the distillate (pH 4.7 to 6.0). For this reason, use of condensate for cooling the compressor lobes had to be discontinued. Approximately 20 gal/hr of fresh DI water was used for this purpose. Moreover, corrosion of the compressor resulted in an iron contamination of the condensate, which had to be disposed of. As a net result, instead of recovering 30 gal of rinsewater per hour, the unit produced 50 gal of wastewater per hour. Using a more expensive vacuum pump made of corrosion-resistant alloys or using a non-sealed compressor might have solved the corrosion problems.

The second problem attributable to the aggressive nature of the condensate water was corrosion of the cast iron distillate pump, which further contaminated the condensate. Replacing this pump with one made of stainless steel reduced iron contamination of the condensate. Nevertheless, the produced water was unsuitable for reuse because of its low pH and had to be treated in the industrial wastewater treatment plant. Additional problems were caused by failure of seals in the concentrated chromic acid pump. These problems were solved by replacing the pump seals and installing new Teflon gaskets. After contamination of the concentrated plating solution had been eliminated, the concentrate from the evaporator was found to be acceptable for reuse in the plating tanks.

Operation and maintenance (O&M) of the LICON unit for evaporative recovery required approximately 2 hours per shift, or 6 hours per day—the equivalent of 1,400 work-hours per year. The mean time between failure (MTBF) was determined to be approximately 40 hours. It should be noted that as the major problems already mentioned were resolved, the O&M effort decreased and the MTBF increased. Regardless, it can be concluded that operation and maintenance of the LICON unit is labor- and capital-intensive.

Table 5.5 is a comparison between the cost of utilizing the LICON unit for evaporative chromium recovery and the cost of operating the chrome plating system without chromium recovery. This comparison includes only those costs affected by the process modification, not the total cost of either operation.

Table 5.5.   Relative Costs of LICON Unit for Chromium Recovery

Cost Type	Description	Without LICON	With LICON
Capital Costs		($ Invested)	
Equipment	LICON system	0	65,800
	Building	0	3,700
	Extra tank	0	3,100
Installation	Building (3MD $200)	0	600
	Installation (10MD $200)	0	2,000
TOTAL CAPITAL COSTS		0	75,200
Annual Costs		**($/Year)**	
Materials	Chromic acid	770	0
	Deionized water[a]	840	230
	Cation regen. chemicals	0	60
	Contaminated DI water treat	560	117
	Maintenance supplies	0	600
Energy	Electricity	0	4,200
O&M Labor	Operation (888 MH $8.00)	0	7,100
	Maintenance (525 MH $8.00)	0	4,200
Disposal	Sludge disposal	1,800	120
Capital Recovery Costs (5 Years, 10%)		0	19,838
RELATIVE ANNUAL COSTS		3,970	36,465

[a]Without LICON, for plating bath makeup; with LICON, for cooling compressor.

This analysis shows that use of the LICON unit for chromium recovery from hard-chromium plating rinsewater is not cost-effective; it adds approximately $30,000 per year to the cost of the existing operation. Moreover, by using flow control and maintaining the recommended maximum contaminant level in the rinse system, the Charleston facility has been able to reduce rinsewater flows to below that required to make up for the water lost by natural evaporation.

Rinsewater cleanup using cation exchange. Since flow control was used to reduce the rinsewater flow to that required to make up for evaporation in the plating tanks, the feasibility of recovering the chromium in the unconcentrated rinsewater by direct addition to the plating tanks was investigated. Since this closed cycle operation could result in a buildup of contaminants, the effectiveness of using the LICON cation exchange unit to remove cationic contamination from the rinsewater prior to reuse was evaluated.

To further reduce the flow of rinsewater required, the cation-exchanged rinsewater was used in spray rinse nozzles located above each plating tank. The recycled rinsewater was found to be contaminated by sulfate, an anion, because platers rinsed parts from an etching tank in the chrome rinse system. Since this practice was discontinued, the system has been working effectively.

Table 5.6 details the costs of recycling rinsewater using cation exchange polishing compared with the costs of chemical makeup, treatment, and disposal of the rinsewater. Recycling rinsewater with cation exchange polishing was found to have a payback period of approximately 5 years.

Table 5.6. Relative Costs of Using Cation Exchange for Chromium Recovery from Rinsewater

Cost Type Capital Costs	Description	Without LICON ($ Invested)	With LICON ($ Invested)
Equipment	Cation exchange unit	0	5,600
	Support equipment	0	2,100
Installation	Labor	0	1,000
TOTAL CAPITAL COSTS		0	8,700

Annual Costs		($/Year)	
Materials	Plating chemicals (makeup)	770	0
	Deionized water (makeup)	840	0
	Cation regen. chemicals	0	170
	Chemical destruction	560	60
Energy	Electricity	0	200
O&M Labor	Operation and maintenance	0	1,000
Disposal	Sludge disposal	1,800	120
Capital Recovery Costs (5 Years, 10%)		0	2,295
RELATIVE ANNUAL COSTS		3,970	3,845

The report claimed that additional savings could be realized if the cation exchange were used at facilities where plating baths are dumped frequently. It should be noted that cation exchange is not feasible for cleanup of concentrated chromium baths, and use on recycled rinsewater only removes those cationic contaminants that would have been removed by drag-out if the rinsewater were not recycled.

Mixed rinse concentration for disposal. In addition to the LICON unit being tested for recovery of chromium from rinsewater, the vapor recompression unit was evaluated for its effectiveness in concentrating a mixture of chromium-containing rinsewaters to reduce the volume that had to be disposed of. Rinsewaters tested were from passivation, anodic stripping, chrome and nickel stripping, copper stripping, chromate conversion, and electropolishing. The vapor recompression evaporator concentrated these rinsewaters to approximately 10% dissolved solids. Problems encountered during the test included scaling of heat transfer surfaces and precipitation of solids in the concentration tank. These are problems that, if left unresolved, could potentially render the process totally ineffective.

Recently promulgated RCRA amendments severely limit the disposal of liquid hazardous wastes. Vapor recompression evaporation does not produce a solid waste, and even with further treatment of this concentrate, the volume of solidified waste would not be significantly less than that produced by conventional treatment. This experiment was an attempt to find a use for the LICON unit, rather than an evaluation of its widespread applicability.

Use of the LICON unit for chromium plating bath cleanup. The feasibility of using the LICON unit to clean up contaminated chromium plating solutions was evaluated. Waste plating solutions were first diluted with water to protect the ion exchange resins, cations were removed in the cation exchange cleanup module, and the solutions were then concentrated to bath strength in the vapor recompression module. These reconstituted baths successfully passed plating tests.

Conclusions. The LICON unit is complicated; it requires close supervision and frequent and expensive maintenance. Its use at facilities with plating shops similar to those at both Charleston and Pensacola is not recommended; low drag-out rates reduce the amount of available chromium in the rinsewater, and more cost-effective and dependable methods to recycle rinsewater and clean up plating solutions are available.

Reverse Osmosis

Reverse osmosis (RO) is a demineralization process in which water is separated from dissolved metal salts by forcing the water through a semipermeable membrane at high pressures (400 to 800 pounds per square inch gauge [psig]). The basic components of an RO unit are a membrane, a membrane support structure, a containing vessel, and a high-pressure pump (Figure 5.11). A typical RO recovery process is

Figure 5.11.   Reverse osmosis equipment (Photo courtesy of Osmonics, Minnetonka, Minnesota.)

shown in Figure 5.12. Rinsewaters must be filtered to prevent fouling the membranes by solid particles. RO units can concentrate most divalent metals (e.g., Ni, Cu, Cd, Zn) from rinsewaters to a 10% to 20% solution. The concentrated solution is fed back to the plating bath to make up for plating and drag-out losses. Activated carbon adsorption is commonly used to remove organic contaminants. The cleaned rinsewater can then be reused.

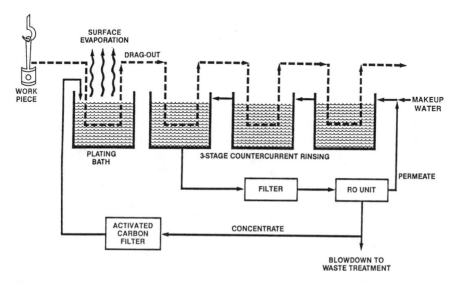

Figure 5.12. Reverse osmosis metal recovery system.

For a typical RO system, capital costs were reported to be $20,000 and annual operating costs were $5,000.[6] As a result of savings associated with plating chemicals, wastewater treatment, and sludge disposal, the payback period for this process modification was reportedly 4.3 years.

According to an EPA study, the main plating application of RO has been for concentrating rinsewaters from slightly acidic nickel plating baths using cellulose acetate membranes.[11] Since 1970, more than 150 RO systems have been installed for nickel plating baths. Recovery efficiencies have been reported between 90% and 95%, with membrane lives ranging from 1 to 3 years.[12]

Approximately 20 RO systems have been installed for recovering copper sulfate, copper cyanide, zinc sulfate, brass cyanide, and hexavalent chromium. RO use for these baths is limited, since RO membranes are attacked by solutions with a high-oxidation potential (e.g., chromic acid) or extremes of pH (less than 2.5 or greater than 11.0). The use of RO for non-nickel baths is expected to increase because of the anticipated development of membranes that can withstand corrosive and oxidizing environments.

RO use is limited to moderately concentrated rinsewaters. For this reason, it is often coupled with a small evaporator when used to concentrate rinsewaters from ambient temperature baths, such as copper and zinc sulfate. An EPA study evaluated the use of RO and evaporation for recovering zinc cyanide from rinsewaters.[13] To reach an adequate concentration for reuse in the ambient temperature plating bath, an evaporator was required to supplement the RO system. Capital costs for the RO system and evaporator were $25,000 and $40,000, respectively, for a total cost of $65,000 (1981 dollars). The operating cost of the complete system was $12,000 per year. A savings of $10,000 per year in wastewater treatment, water, and makeup chemical costs was insufficient to offset operating and capital recovery costs.

Another EPA study[14] demonstrated that RO could be effectively used to recover copper cyanide from rinsewater for recycling in a plating bath. However, because of low rinsewater concentrations, short membrane lives, and low wastewater disposal costs, this process was found not to be cost-effective.

In summary, RO has been shown to be cost-effective in concentrating nickel in rinsewaters for reuse in nickel plating baths. However, for ambient temperature plating baths, RO must be supplemented with expensive evaporators to concentrate the metals in rinsewater to plating bath strength. The cost-effectiveness of an RO metal recovery system depends on production rate, type and concentration of constituents in the rinsewater, costs of fresh water supply and wastewater disposal, and expected useful life of the RO membrane used. Process and operating uncertainties associated with membrane processes that can significantly affect their cost-effectiveness include membrane fouling, bath chemical balance, wastewater generation, and operation and maintenance requirements. Suppliers of RO equipment are listed in Table 5.7.

Table 5.7. Suppliers of Reverse Osmosis Equipment

Company	Address	Phone
Osmonics	5951 Clearwater Drive Minnetonka, MN 55343	(612) 933-2277
Separation Technology	454 S. Anderson Rd. Rock Hill, SC 29703	(803) 329-1252

Electrodialysis

Electrodialysis (ED) concentrates or separates ionic species in a water solution through use of an electric field and semipermeable ion-selective membranes. Applying an electrical potential across a solution causes migration of cations toward the negative electrode and migration of anions toward the positive electrode. ED units are packed with alternating cation and anion membranes. Cation membranes pass only cations, such as copper, nickel, and zinc, whereas anion membranes pass only anions, such as sulfates, chlorides, or cyanides. Alternating cells of concentrated and dilute solutions are formed between the cation and anion membranes. Packaged ED units contain from 10 to 100 cells.

Electrodialysis has been used to recover cationic metals from plating rinsewaters. In a typical application, as depicted in Figure 5.13, rinsewater from a stagnant or dead rinse (i.e., no inflow or outflow) tank is continuously fed to an ED unit and concentrated by a factor of 10. The concentrate is then returned to the plating bath. The waters in the dilute cells are combined with makeup water and returned to the dead rinse tank.

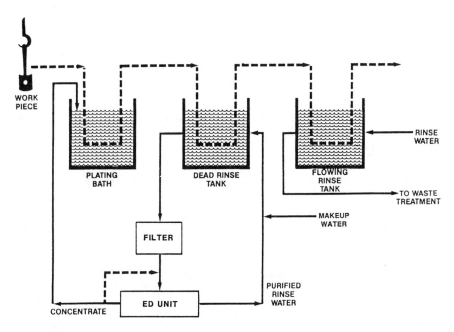

Figure 5.13. Electrodialysis metal recovery system.

Unlike ion exchange (IX) and reverse osmosis (RO), the maximum concentration of an ED unit is limited only by the solubility of the compounds in solution. Therefore, ED generally can produce a more concentrated solution than IX and RO, eliminating the need for an evaporative concentrator when used with ambient-temperature plating baths. ED units are reportedly easy and economical to operate, require little space, and operate continuously without requiring regeneration.[3]

One disadvantage of ED and RO is that all ionic species are nonselectively removed. Therefore, ionic impurities—organic brighteners, wetting agents, and other nonionized compounds that accumulate in the dead rinse tank—are returned to the plating bath along with the recovered metal. As a result, plating baths must be periodically treated to remove impurities, and the dead rinse tanks must occasionally be disposed of. Also, if the applied voltage exceeds the hydrogen electrode potential, water will be converted to gaseous hydrogen and hydroxide ions. The subsequent increase in pH can cause precipitation of metal hydroxides that can foul the membranes.[15]

ED package systems cost from $30,000 to $45,000 (1984 dollars). A Navy study estimated a payback period of less than one year for an ED recovery system for a cadmium cyanide plating bath operating 4,000 hours per year at drag-out rates of 1.3 lb/day of Cd and 5.1 lb/day of CN. This evaluation did not include the costs of removing impurities from the baths or of maintaining the ED units and replacing the membrane modules.[3]

An EPA study evaluated recovery of nickel from rinsewaters using ED.[16] The ED unit was able to recover 95% of the nickel salts from the rinsewater and return the concentrated solution to a Watts-type nickel plating tank. The study estimated that $16,000 per year could be saved by employing ED in a nickel plating line that operated 4,000 hours per year. The cost estimate considered only savings in chemical usage, wastewater treatment, and sludge disposal and did not consider the cost of operating and maintaining the ED system.

The following case study illustrates the use of innovative rinsing to recover plating solutions from rinsewaters and the use of electrodialysis to remove contaminants from recycled plating solution.

Case Study 5.2: Recovery of Chromium by Innovative Rinsing Techniques

The Naval Aviation Depot (NADEP) at Pensacola is a government-owned, government-operated (GOGO) facility employing approximately 4,000 people. The primary mission of the facility is to recondition H-3 and H-53 helicopters and A-4 jet aircraft. Reconditioning consists of disassembling the aircraft and components; stripping paint; removing dirt, grease, and corrosion products; remanufacturing or replacing parts; reassembling parts; and applying protective coatings (plating and painting). When replacing worn parts with new ones is not feasible, parts are remanufactured by overplating with chromium (hard-chrome plating), followed by machining back to original specifications.

The most common electroplating process found at NADEPs, Naval Shipyards (NSYs), and Naval Air Stations (NASs) is hard-chrome plating. Hard-chrome plating methods used at Naval facilities have remained essentially unchanged for more than 20 years, despite advancements in plating technology and concerns with environmental pollution. Areas on worn parts that do not require a chrome buildup are masked with wax, aluminum foil, lacquer, or tape. After masking, the parts are fastened to racks and suspended in the plating bath. These racks are then secured by C-clamps to the cathode bus bar, which provides physical support for the part and completes the electrical circuit. Heavy lead anode bars are then hung from the anode bus bar and positioned around the racked part. Since the lead anodes are 8 ft long and weigh more than 50 lb each, they cannot be easily removed by one person, so they are often left in the plating solution when not in use. As a result, the anodes slowly become passive and ineffective.

After plating, parts must be rinsed to remove plating solution that is dragged out of the bath. Continuous flow rinse tanks are generally used to clean plated parts. Rinse flows range from 3 to 12 gal/min, resulting in a cost of $7,000 to $28,000 per year per rinse tank at Pensacola. This figure is based on 24 hr/day, 260 days/yr of operation (a freshwater cost of $0.34/1,000 gal and a wastewater treatment cost of $5.81/1,000 gal).

Facilities for hard-chrome plating require large production areas; only one or two large parts can be plated at the same time in a single tank, and plating times often exceed 24 hr. The period between receipt of a part at the plating shop and delivery to the machine shop is often a week or more. Since these parts are frequently critical to repairing an aircraft, plating delays can significantly extend time for maintenance.

Hard-chrome plating is considered the most demanding of all plating processes because it requires close supervision and a high degree of quality control. Most parts require a uniform buildup of chrome so they can be accurately ground and polished to their required dimensions. Platers using current Navy plating methods have had trouble meeting these specifications and quality requirements. These methods often result in uneven plating deposits because the anodes cannot be arranged to provide a uniform current density on the surface of parts. Rejection rates have been as high as 40%. Rejected parts are stripped and returned for replating, causing an increased workload for the plating and machine shops and delays for delivery of the remanufactured part.

Plating baths become contaminated with metal ions leached from parts, plating tanks, racks, and anodes, and with conversion of hexavalent to trivalent chromium. These contaminants can blemish a plated surface, reducing plating efficiency and quality. When baths are deemed unsuitable for use, they are bled into the industrial waste system. Because of a buildup of impurities, plating baths at Pensacola have been dumped about every 2 years. Approximately three times a year, plating baths have been accidentally discharged to the sewer because the plating tanks were not equipped with high-level alarms. Treating plating wastewater and replacing the plating solution with new material is expensive.

To mitigate these problems, the Naval Civil Engineering Laboratory (NCEL) at Port Hueneme, California, adapted an innovative chrome plating system for use at Navy plating shops.[3] The "new" plating process uses a technology that was developed more than 50 years ago in the Cleveland area, hence it is called the "Cleveland process" or the "Reversible Rack 2 Bus Bar System." The Navy laboratory converted three of the seven plating baths at Pensacola to the Cleveland process to demonstrate this technology. Approximately 50% of hard-chrome plating at Pensacola now uses this innovative system. Although the plating method varies considerably from conventional procedures, the system greatly improved plating efficiencies, and the end product meets all military specifications.

Modifications from the standard Navy practice of hard-chromium plating were as follows:

1. use of conforming anodes and reversible racks to suspend parts instead of the Navy practice of clamping parts on a third cathodic bus bar and using common lead anodes

2. control by voltage (4.5 volts) rather than by amperage

3. use of a recirculating spray rinse system (Figure 5.14)

4. operation at higher temperatures (140°F vs 130°F)

5. use of a continuous system of bath purification to remove contaminating cations from the plating solution (Figure 5.15)

Photographs of the components of this innovative chrome plating system are shown as Figures 5.16 through 5.21.

Use of conforming anodes has produced a more even current density for the Cleveland process. The process provides a more uniform deposit, an improved product quality, and an increased plating rate. The reversible racks require considerably less room in the plating tank than the conventional system of clamping anodes and parts to three bus bars. Also, control of the process is greatly simplified by using voltage rather than amperage, and multiple parts can be plated in the same tank simultaneously.

To ensure good adhesion of new plating to the existing surface, parts are often subjected to a reverse current to etch or roughen the existing surface. The conventional Navy process requires an expensive switching mechanism to reverse polarity of the bus bars for etching. All of the parts in a tank can be either plated or etched at any one time, but both operations cannot be performed concurrently. In contrast, the Cleveland process uses reversible racks that can be picked up and placed in the

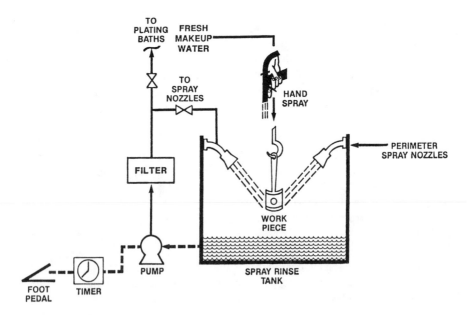

Figure 5.14. Recirculating spray rinse system.

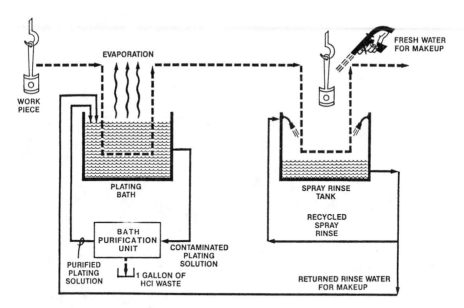

Figure 5.15. "Zero-discharge" chrome plating system.

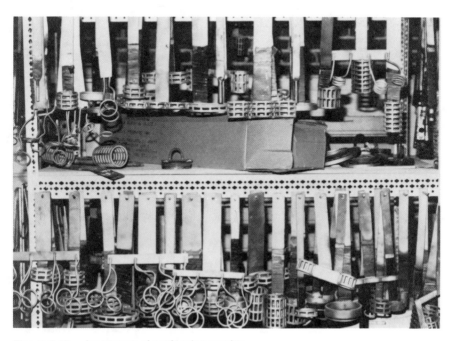

Figure 5.16. Assortment of conforming anodes.

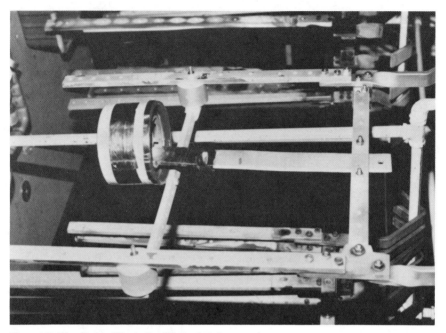

Figure 5.17. Reversible rack with conforming anode.

Figure 5.18. Chrome plating bath with two bus bar reversible rack.

Figure 5.19. Spray rinse of plated part and reversible rack.

opposite direction to reverse current for etching. With this method, some parts in a tank can be plated while others are being etched. In addition, 16 to 20 parts of different sizes and shapes could typically be plated simultaneously in one tank, compared to only 6 to 8 using the conventional Navy system.

To reduce the amount of rinsewater used, a prototype spray rinse system was installed in an existing rinse tank (Figure 5.14). A foot-activated pump recirculates rinsewater through eight high-velocity spray nozzles located around the perimeter of the rinse tank. Clean rinsewater is also available from a handheld sprayer. After repeated use, a portion of the rinsewater is pumped through a cloth filter into the plating tank to replace water lost to evaporation (Figure 5.15). When compared to conventional chrome plating, operation of the plating bath at a higher temperature results in higher evaporation and increased plating rates. Because of these changes, less rinsewater is produced than is needed to make up for evaporation losses in the plating bath. The result is a "zero discharge" plating system.

Without drag-out to help remove contaminants from the bath, a cleanup process was desirable to reduce the need for plating bath dumps. A "CatNapper-10" treatment system, manufactured by Innova Technology, Inc., of Clearwater, Florida, was installed to continuously remove metal cations from the chromium plating bath. According to the vendor, the CatNapper system uses a cathode contained within a membrane module. The cathode selectively precipitates contaminating metal cations from the plating solution and oxidizes trivalent chromium to its hexavalent form.

Figure 5.20. Hand spray rinse.

Figure 5.21. "CatNapper 10" electrolytic plating bath purification system.

Hexavalent chromium remains on the anode side of the membrane and is returned to the plating bath. Because it removes impurities from the plating baths, the Cat-Napper is supposed to extend the life of the bath and reduce the need for increased chrome concentrations or plating voltage. By extending the life of a bath (and thus decreasing the frequency of bath dumps), the CatNapper could indirectly reduce the volume of hazardous waste produced by the plating shop.

The cost and savings associated with this modification are summarized in Table 5.8. The savings result principally from reduced wastewater treatment costs. Of this savings, approximately $125 is attributable to recovering chromium. Innova, the

Table 5.8. Capital Costs for Innovative Chromium Plating System

Item	Cost
Convert plating tanks equipment, materials, labor	$ 5,950
Subtotal	$ 5,950
Bath purification system CatNapper & rectifier	$ 8,900
Misc. equipment	$ 1,460
Labor	$ 1,000
Subtotal	$11,360
Spray rinse system	
Equipment	$ 2,140
Labor	$ 1,800
Subtotal	$ 3,940
TOTAL INSTALLED COST	$21,250
Annual cost reduction (waste treatment)	$25,000
Payback period (years)	0.9

manufacturer of CatNapper, is reportedly no longer in business. A similar but improved system is currently supplied by IONSEP Corporation. The IONSEP process consists of a membrane electrochemical cell designed for immersion in a process liquor, a rectifier, a process liquor containing metal salts, an IONSEP catholyte solution, and a pump to flow the catholyte solution through the cell. The membrane separates the process liquor from the catholyte solution and acts as an "electrochemical traffic controller" that lets metal cations go from the process liquor through the membrane (electrofilters the metals) into the catholyte solution and keeps anions in the process liquor. The metal cations are continuously converted to hydroxides in the catholyte solution and the anions are continuously converted to acids in the process liquor. The hydroxides of multivalent metals (cadmium, zinc, iron, copper, aluminum, calcium, etc.) are substantially insoluble in the catholyte and can be removed for use. The IONSEP process is unique in that salts of multivalent metal cations can be converted. There is essentially no electrodeposition of metals. The

IONSEP process can be operated at reproducible capacities for months. The capacity of the IONSEP process is varied by voltage (Figure 5.22).

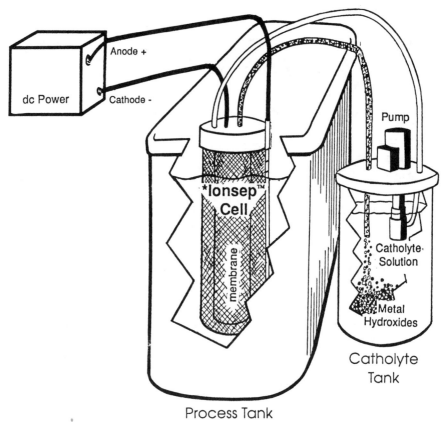

Figure 5.22.   Schematic of electrodialytic bath cleanup system (Diagram courtesy of Ionsep Corporation, Inc., Rockland, Delaware).

Reducing or Eliminating Tank Dumping

For an operation to maintain a "zero discharge" or "closed loop" system, it is necessary to periodically remove impurities from the plating bath (Figure 5.13). Divalent metal impurities such as iron and trivalent chromium need to be removed from chromium plating baths. Nickel baths are usually purified by activated carbon adsorption. Carbonates, the principal impurities in cyanide baths, are normally removed by chemical precipitation.[17] Suspended solids are removed by cartridge filtration.

Substituting Less Hazardous Materials

Noncyanide Baths

Noncyanide baths have the advantages of improving workplace safety, and reducing rinsewater disposal costs. Traditionally, cadmium, zinc, brass, and precious metals have almost universally been plated from alkaline cyanide baths because of the superior plate produced from stable metal cyanide complexes. Unfortunately, cyanide baths are costly and dangerous to operate, and the wastes generated are difficult and expensive to treat.

In the late 1960s and early 1970s, extensive research was conducted to develop noncyanide zinc electroplating baths. As a result, several alternative zinc baths were developed. Alternatives include low-cyanide baths; noncyanide alkaline baths; neutral ammonium chloride and potassium baths; and a number of acidic baths containing sulfate, chloride, and fluoborate ions.[3]

Low-cyanide baths contain approximately 20% as much cyanide as conventional cyanide baths and have similar operating characteristics. However, process control is more difficult, and cyanide treatment is still required.

Neutral chloride baths use ammonium or potassium ions to form a zinc complex. These baths usually require the addition of proprietary brighteners and chelating agents that form zinc complexes. Unfortunately, zinc complexes can be difficult to remove in subsequent waste treatment.

Acid sulfate, chloride, and fluoborate baths have become the most popular noncyanide zinc baths. With the recent development of new additives, acid zinc baths are capable of producing bright deposits that are competitive with alkaline cyanide baths for general plating applications.[3]

Less effort has been expended in developing noncyanide cadmium baths because the volume of cadmium plating is only 5% to 10% that of zinc plating. However, because of increased environmental and safety concerns with operating and disposing of cadmium cyanide baths, alternative proprietary acidic cadmium baths similar to zinc baths have recently been developed to replace cyanide baths.

Most of these acidic baths consist of cadmium oxide, sulfuric acid, distilled water, and anion compounds. Because many old tanks for alkaline cadmium cyanide plating are made of bare steel, conversion to these acidic baths may require that the existing tanks be refurbished or replaced. Thus, material substitution may require a considerable capital expenditure. However, the savings in eliminating cyanide treatment can make the modification economically attractive.[18]

Parts being plated in noncyanide cadmium baths may require more thorough cleaning before plating than parts being plated in cyanide baths. The noncyanide cadmium baths reportedly have less throwing power and lower cathode efficiency than cyanide baths. Despite the disadvantages, however, some platers prefer the new noncyanide plating baths because of the reduced complexity of waste treatment. Some platers have reported that drag-out of noncyanide cadmium baths is less than that of cyanide baths.

Noncyanide zinc and cadmium baths usually cost more than cyanide baths. However, to properly evaluate the cost-effectiveness of the material substitution, facilities must also consider the following factors: cost of new corrosion-resistant equipment, difference in labor and chemical costs, change in production rate, and savings realized by eliminating cyanide treatment.

Example. In 1983, the Charleston Naval Shipyard successfully switched from alkaline cyanide baths to an acidic noncyanide solution and eliminated the cyanide oxidation process from the waste treatment plant.[19]

Example. The plating shop of Air Force Plant #6, located in Marietta, Georgia, and operated by Lockheed-Georgia Corporation, switched from an alkaline cyanide cadmium bath to a proprietary acidic noncyanide cadmium bath. Lockheed found that product quality was improved by switching from the alkaline cyanide baths to the acidic noncyanide cadmium baths; however, more careful process control was required. The new plating solution, costing approximately $3 per gal, is more expensive than the old formulation; however, reduced costs for waste treatment resulted in a net savings, since the wastewater treatment plant no longer had to operate the treatment process for alkaline chlorination cyanide. The material substitution was implemented primarily to reduce the safety hazards associated with the operation and disposal of cyanide baths. Improved quality and decreased costs have ensured the permanent adoption of the process modification.

Trivalent Chromium Plating

Trivalent chromium baths have been used in place of conventional solutions of hexavalent chromium. With trivalent chromium rinsewaters, it is unnecessary to add sodium bisulfite or other reducing agents in waste treatment for conversion of hexavalent to trivalent chromium before precipitation. Trivalent solutions are also less concentrated (22 g/L versus 150 g/L for hexavalent solutions), thus lessening the amount of chromium drag-out. Consequently, sludge produced from trivalent baths is about one-seventh of the volume from hexavalent baths and is much less toxic.[20]

The main disadvantage of trivalent solutions is that they cost two to three times more than hexavalent solutions. Some researchers have reported that higher production rates and lower rejection rates can be realized with trivalent chromium plating solutions; however, the main advantage is the lower cost of wastewater treatment and sludge disposal. Before a plating shop converts to trivalent chrome solutions, a detailed study should be performed to determine if the projected savings in waste treatment exceed the increased operating costs.

Electroless Nickel Plating

Electroless nickel plating was developed in 1946 to coat a substrate without using an outside source of electrical current. Electroless nickel plating uses the substrate to catalyze a chemical reduction reaction. However, because of the expense of the chemical reducing agents, electroless plating is not cost-effective in applications where conventional electroplating can be used.

The majority of nickel plating is done in an acidic (pH between 1.5 and 4.5), elevated-temperature (between 110°F and 150°F) Watts bath that contains nickel sulfate, nickel chloride, and boric acid. An electrical current causes the nickel to be plated on the substrate.

The main advantages of electroless nickel plating are that the throwing power is essentially perfect and the deposits provide greater protection of the substrate because they are less porous.[3] In addition, the nickel concentrations of electroless baths are approximately one-eleventh those of conventional Watts nickel baths. Therefore, drag-out quantities and sludge production from an electroless bath are much less than from a conventional bath.

Vacuum Deposition of Cadmium

Vacuum deposition of cadmium was developed as an alternative to electroplating. Problems with electroplating arise from cadmium cyanide baths because of the toxicities of cadmium and cyanide. Switching to noncyanide plating baths (discussed previously) removes one of these problems. Use of vacuum deposition of cadmium also eliminates the need for cyanide.

Vacuum deposition of cadmium is a line-of-sight process, making it difficult to provide a uniform deposit on an irregularly shaped part. Parts need to be rotated at intervals during processing for more uniform coverage. Adhesion of the deposit to the base metal is not as strong as that produced by conventional cadmium plating. Also, occupational and environmental hazards can result from the evacuation of cadmium vapors and condensed aerosols. In addition, the vacuum exhaust must be carefully filtered to prevent these cadmium vapors and condensed aerosols from escaping to the work environment.

Ion Vapor Deposition of Aluminum

Ion vapor deposition of aluminum is a substitute for cadmium plating for corrosion protection. It was developed because of the many hazards inherent in working with cadmium and the increasingly stringent requirements being placed on disposal of wastes that contain even traces of cadmium. Aluminum coating is a logical replacement for cadmium to provide corrosion protection because aluminum is anodic to steel and provides galvanic protection similar to that afforded by cadmium. In addition, aluminum's corrosion products are not bulky or unsightly. Aluminum is also less expensive than cadmium and zinc on a volume basis. Moreover, aluminum can be used at temperatures up to 925°F, compared with a maximum of 450°F for cadmium.

As a result, there has been considerable interest in the possibility of aluminum plating, with many attempts to develop a successful method. However, the electrode potential of aluminum is too negative for it to be successfully plated from an aqueous solution.[21] Aluminum has been deposited on steel by hot dipping or by using a metal spray system. These methods do not provide the thin, uniform coating required on aircraft parts, nor do these coatings adhere to substrates as strongly as plated cadmium.

As a logical extension of vacuum deposition of cadmium, ion vapor deposition (IVD) of aluminum was developed by McDonnell Douglas Corporation as a substitute for cadmium plating on steel aircraft parts.[22-24] The IVD (Ivadizer) system (Figure 5.23) consists of a vacuum chamber, a resistance-heating aluminum-vaporization system, and a high-voltage system to ionize the aluminum and to impart a negative charge to the parts. The charge causes aluminum ions to electrodeposit on the parts. Air in the vacuum chamber is replaced by a low-pressure inert gas, which is ionized. Aluminum vapor must interact with the ionized inert gas for the aluminum to be ionized and attracted to the oppositely charged parts and to coat them uniformly. Without this ionization and interaction with the inert gas ions, IVD would be restricted to line-of-sight coating as in the vacuum deposition of cadmium.

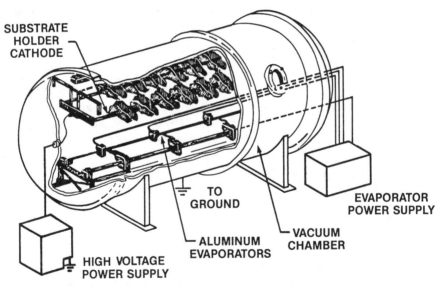

Figure 5.23. Equipment for ion vapor deposition of aluminum.

Advantages cited for IVD of aluminum include a higher useful temperature, improved throwing power, and better adhesion of the aluminum coating compared to cadmium. In addition, parts that are cadmium-plated require baking to prevent hydrogen embrittlement; problems have been encountered with oven temperatures not being carefully controlled, so that parts must be scrapped. Safer working conditions are cited as another advantage of IVD of aluminum.

Example. Personnel at the North Island Naval Aviation Depot (NADEP) have been using IVD of aluminum for about 11 years, having procured one of the first commercially available systems. After many problems with this developmental model, NADEP procured a more recent, improved model. However, numerous problems persist in adapting the process to existing plating operations.

Problems developed with the newer IVD system for two principal reasons. First, the equipment was installed in the open plating shop, where considerable contamination of the vacuum chamber by ambient gases resulted. Second, the IVD system is complicated to operate because many operating variables need to be adjusted to produce a good coating. Personnel from the plating facility were assigned to operate this complicated technology without adequate skills, training, or incentives.

Production personnel oppose complete conversion to IVD because the process is more complex and requires more labor and skill than cadmium plating. For this reason, parts are being evaluated on an individual basis for conversion to IVD aluminum coating. Most parts are still either electroplated from cadmium cyanide baths or vacuum-deposited with cadmium. The limited use of IVD has aggravated problems with the system at the depot, because extensive use is needed to, as one facilities engineer put it, "work the bugs out of the system." This limited use of the system and a general lack of cost information by personnel made it impossible to evaluate the economics of the process.

In summary, the technology appears to have considerable potential. When the process is performed correctly, the coating is as protective as cadmium coating. From an environmental standpoint, widespread adoption to replace cadmium plating would eliminate a significant source of hazardous waste. However, unless these systems are made easier to operate and maintain, unless they are located in cleaner facilities than plating shops, and unless they are supported by skilled and well-trained operators, it is unlikely that IVD of aluminum will displace cadmium plating at NADEPs.

REFERENCES

1. Johnnie, S. T. "Waste Reduction in the Hewlett-Packard, Colorado Springs Division, Printed Circuit Board Manufacturing Shop," *Hazardous Waste & Hazardous Materials.* 4(9) 1987.
2. "Control and Treatment Alternatives for the Metal Finishing Industry, In-Plant Changes." U.S. Environmental Protection Agency, EPA 625/8-82-008 (January 1982).
3. Cushnie, G. C. "Navy Electroplating Pollution Control Technology Manual," written for Naval Civil Engineering Laboratory, Port Hueneme, CA, Report No. 84.019 (February 1984).
4. "Environmental Pollution Control Alternatives: Economics of Wastewater Treatment— Alternatives for the Electroplating Industry," U.S. Environmental Protection Agency, EPA 625/5-79-016 (June 1979).
5. Moore, Gardner & Associates. "Naval Shipyards Industrial Process and Waste Management Investigation," prepared for Naval Facilities Engineering Command, Contract No. N00025-80-C-0015 (July 1983).

6. Mouchahoir, G. E., and M. A. Muradaz. "Clean Technologies in Industrial Sectors of Metal Finishing, Non-Ferrous Metals, and High Volume Organic Chemicals," U.S. Environmental Protection Agency, EPA 68-01-5-21 (June 1981).

7. "Summary Report—Control and Treatment Technology for Metal Finishing Industry—Ion Exchange," developed by the Industrial Environmental Research Laboratory, EPA 625/8-81-007 (June 1981).

8. "Water Pollution Abatement Technology Capabilities and Costs: Metal Finishing Industry," Lancy Laboratories, NTIS PB-24 808 (October 1975).

9. Carpenter, C. J. "Test and Evaluation of the LICON Chrome Recover Unit at NARF Pensacola," Technical Note No. 2845TN, Naval Civil Engineering Laboratory, Port Hueneme, CA (January 1984).

10. Mordecai, B. S., and G. C. Bradley. "Testing and Evaluation of a Chrome Recovery System Utilizing a High Vacuum Vapor Recompression Evaporator and a Cation Exchange Module at the Charleston Naval Shipyard," a report to the Environmental Branch, Utilities Division, Southern Engineering Field Division, Naval Facilities Engineering Command, Charleston, SC (April 1985).

11. McNulty, K. J., and P. R. Hoover. "Evaluation of Reverse Osmosis Membranes for Treatment of Electroplating Rinsewater," EPA-600/2-80-084, NTIS PB80-202385 (May 1980).

12. Cartwright, P. S. "An Update on Reverse Osmosis for Metal Finishing," *Plating and Surface Finishing*, April 1984.

13. McNulty, K. J., and J. W. Kubarewicz. "Demonstration of Zinc Cyanide Recovery Using Reverse Osmosis and Evaporation," EPA-600/2-81-132, NTIS PB-231243 (July 1981).

14. McNulty, K. J., et al. "Reverse Osmosis Field Test: Treatment of Copper Cyanide Rinse Waters," EPA-600/2-77-170, NTIS PB-272473 (1979).

15. Eisenmann, J. L. "Membrane Processes for Metal Recovery from Electroplating Rinse Water," Second Conference on Advanced Pollution Control for the Metal Finishing Industry, EPA-600/8-79-014, pp. 99-105 (June 1979).

16. Eisenmann, J. L. "Nickel Recovery from Electroplating Rinsewaters by Electrodialysis," NTIS PB81-227209, EPA-600/2-81-130, July 1981.

17. Hartley, H. S. "Evaporative Recovery in Electroplating," First Annual Conference on Advanced Pollution Control for the Metal Finishing Industry, EPA-600/8-78-010, pp. 86-91 (May 1978).

18. Jorczyk, E. R. "A New Non-Cyanide Cadmium Electroplating Bath. Water Pollution Abatement Technology Capabilities and Costs: Metal Finishing Industry," prepared for National Commission on Water Quality, NTIS PB-248 808, (October 1975).

19. Cushnie, G. C. "Initiation Decision Report—Treatment of Electroplating Wastes," prepared for Naval Civil Engineering Laboratory, TM No: 54-83-20CR (October 1983).

20. Garner, H. R. "Meeting the Regs: How Trivalent Helps," *Products Finishing* 47(12) (September 1983).

21. Lowenheim, F. A. *Electroplating*. (New York: McGraw-Hill, 1978).

22. Muehlberger, D. E. "Ion Vapor Deposition of Aluminum: More than Just a Cadmium Substitute," *Plating and Surface Finishing* 24 (November 1983).

23. Fannin, E. R. "Ion Vapor Deposited Aluminum Coatings," Proceedings of the Workshop on Alternatives for Cadmium Electroplating in Metal Finishing, EPA-560/2-79-003, p. 68 (1979).

24. Steube, K. E. "Fabrication and Optimization of an Aluminum Ion Vapor Deposition System," Technical Report AFML-TR-78-132, Wright-Patterson Air Force Base, Ohio: Air Force Materials Laboratory (June 1978).

CHAPTER 6

Painting and Coating

DESCRIPTIONS OF PAINTING PROCESSES AND WASTES

Solvent-based paints are applied to surfaces of parts, components, and assembled products for corrosion protection, surface protection, identification, and aesthetic appeal. Most painting is performed by conventional liquid spray technology.[1] In spray painting, the paint is mixed with a carrier, usually an organic solvent, and is applied to the surface with an air-pressurized sprayer (Figure 6.1). Spray painting is usually done in a horizontal or downdraft paint spray booth.

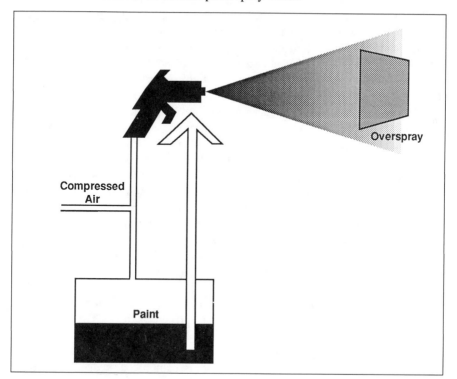

Figure 6.1. Air-atomized spray painting (Source, Reference 3).

Painting processes generate two significant sources of hazardous wastes: paint sludges and waste solvents. The first and largest volume of hazardous waste generated in painting involves air emissions that create paint sludges. During typical spray painting, 50% of the paint is deposited on the surface being painted; the other 50%, called overspray, is sprayed into the air.[2] As the paint dries, the solvent evaporates to the air. The air from a paint booth is often exhausted through a water scrubber that separates the paint from the air. The scrubber water is normally recycled, and paint solids are concentrated in the scrubber sump. When the sump fills with paint sludge, it is removed and put in drums for hazardous waste disposal.

The second significant source of hazardous waste generated in painting processes is the use of solvents to clean painting equipment. Most paints are solvent-based, so they require solvents for cleanup. The type of solvent used varies with the paint. Some of the more common solvents are methyl ethyl ketone (MEK); xylene; 1,1,1-trichloroethane; toluene; butyl acetate; ethylene glycol; monoethyl acetate; and alcohol. In addition, paint stripping solvents (as described in Chapter 7) may be used to remove hardened paint from surface. These wastes are classified as hazardous because of their general flammable and toxic properties.

Many painting processes generate varied types of hazardous waste. The common feature of almost all of these wastes is that their hazardous characteristics are derived both from the paint constituents (heavy metals and solvents) and from the solvents used in cleanup (toxic and flammable organics). For example, when dry scrubber paint booths are used instead of water scrubbers, the filter material can become contaminated with the paint and require disposal as a hazardous waste. However, if water-based paints are used, solvents are not needed for cleanup and thereby the volume of hazardous wastes generated is reduced.

Another environmental discharge from painting is volatile organic compounds (VOCs) emitted to the air. Presently, EPA regulates only VOC emissions from paint coatings and has no regulations regarding solvents used in cleanup. Federal VOC limits for paint are 420 g/L for paints that cure below 90°C and 360 g/L for paints that cure above 90°C.[3] Some state air pollution control agencies are setting strict VOC content limits for paint. For example, the South Coast Air Quality Management District in California has set a 350 g/L VOC limit for aircraft paints.[4] Local regulatory agencies also control VOCs by setting total permissible discharge limits from facilities. These limits include point sources and fugitive sources. The EPA is required to develop limitations for toxic emissions in the future; these limitations will likely have an impact on both the types of solvents used in paint and those used in cleanup.

Most facilities house both paint stripping and painting operations in the same area. Compared to the disposal of wastes from stripping processes, disposal of paint waste is less problematic because much less hazardous waste is generated, and only a small amount of wastewater is produced.

Alternatives to conventional solvent-based spray painting can reduce hazardous pollutants. These alternatives require an integrated approach in which painting techniques are improved and processes are used to reduce or eliminate hazardous materials. For example, modified painting techniques can minimize the amount of wasted paint that ultimately must be disposed of as hazardous waste. Also, companies can use paint formulas that minimize or eliminate solvent paint thinner and cleanup solvents, both of which contribute to hazardous waste and air pollution. In addition, substitute solvents may be used to minimize air pollution and to produce less toxic hazardous waste.

The more promising developments in the area of painting modifications and substitutions are summarized in the following pages.

POWDER COATING TECHNOLOGY

Powder coating, also called "dry powder painting," is one of the major advances in paint application. This technique is based on depositing specially formulated thermoplastic, or thermosetting, heat-fusible powders on metallic substrates. No solvents are used; therefore, the system eliminates the pollution and safety problems associated with solvent-based paints. Powder coating almost completely eliminates air emissions of VOCs, greatly reduces cleanup solvents, and eliminates use of paint thinners. Also, there is no waste (old) paint to dispose of.[5]

In addition to the environmental advantages offered by dry powder painting, the process provides technical, production, and cost benefits. Productivity is increased because, without solvents, the coating can be cured immediately after application. Because curing is thermoactivated, curing times are short. One technical advantage is that special coating materials such as nylon, which cannot be applied by conventional solvent-based systems because of the lack of appropriate solvents, can be applied by dry powder painting. In addition, complex surfaces are more evenly coated in dry powder systems. For some applications, a single coating can replace the multiple-coating applications used in conventional spray painting. Dry powder techniques are also readily adaptable to current production methods and are easily learned by painting personnel.

The one major limitation in dry powder painting is that the items to be painted must be able to withstand the typical curing temperatures of 350°F for 30 minutes.[6] Aluminum alloys cannot be subjected to these conditions without significant loss of strength.

Dry powder painting techniques that are commercially available include electrostatic dry powder painting, fluidized bed method, and plasma spraying. A description of each follows.

Electrostatic Dry Powder Painting

Electrostatic dry powder painting is the most widely used powder coating technique. In this method, dry powder is sprayed onto the surface, where it is electrostatically deposited. The dry powder is metered into a compressed-air-driven spray gun and is sprayed at the surface. An electrode in the spray gun ionizes the air and powder suspension using direct current, and the dry powder particles then become charged. The surface to be coated carries the opposite charge, and the powder is electrostatically attracted to the surface. As the coating builds up, the coating thickness is limited by the loss of attraction of the powder to the surface, resulting in a uniform thickness, even on complex shapes. The coating is then fused to the surface and cured in conventional ovens.

In commercial applications, the painting system collects the powder overspray in conventional air filter systems and reuses it, thus eliminating disposal of overspray water associated with liquid solvent-based paints. A powder use of 90% to 99% is possible. Table 6.1 compares annual operating costs of a dry powder system with a conventional solvent painting system. The comparison is based on operating and maintenance costs required to coat 12 million ft² of parts with a 1-mil polyester coating.[5]

Table 6.1. Cost Comparison of Solvent Painting and Powder Painting

Item	Conventional Solvent	Dry Powder
Material	$333,600	$242,400
Labor and cleanup	132,100	75,600
Maintenance	18,000	10,000
Energy	29,100	15,700
Hazardous waste disposal	10,800	1,100
Total annual cost	$523,600	$344,800
Cost per square foot	$ .044	$ .029

Case Study 6.1: Dry Powder Painting at Hughes Aircraft Company

Developmental program description. At Air Force Plant No. 44, operated by the Missile Systems Group of Hughes Aircraft Company in Tucson, Arizona, electrostatic dry powder painting is being used in a developmental program to paint missile parts. Selected for the initial application of paint to the inside fuselage section of the Phoenix missile, dry powder painting was chosen over other conventional paint systems because of enhanced surface protection, better coverage, and reduction in solvent emission.

The developmental program results have been very successful. In addition to satisfactorily achieving the initial goals, the dry powder painting system developmental program has shown that this system provides additional significant benefits, including reduced hazardous waste, elimination of wastewater, fewer work-hours, less paint use, and lower overall cost per square foot of painted surface. The developmental program is continuing, and in-house implementation is being evaluated.

Industrial process description. Most painted parts used in the fabrication of missiles are painted using the solvent-based wet spray technique. Paint is applied in spray booths, where overspray is collected in a conventional air ventilation system equipped with a recirculating water curtain scrubber, which removes the over-spray from the exhaust air. The scrubber wastewater containing the overspray is treated in the facility's central wastewater treatment system, where the overspray ultimately becomes part of the treatment plant wet sludge, which is a hazardous waste. Waste solvents and paint/solvent wastes are also generated from mixing operations, cleaning operations, empty containers, and waste materials; all are hazardous wastes and require hazardous waste disposal.

Alternative painting technologies were evaluated for painting the interior surface of the fuselage section of the Phoenix missile, an area not previously painted. Coating this area was desirable to enhance corrosion protection from the salt environment in aircraft carriers and from SO_2 in jet engine exhaust. The area is small, approximately 9 ft², and unit production is nominal, approximately 50 per month. The requirements provided the opportunity to test alternative painting technologies on a developmental scale.

Process modification description. The paint system is a polyester and epoxy powder coating that is electrostatically applied and fusion bonded. Paint materials are Type I, thermosetting polyester epoxy powder base coating, or Type II, thermosetting epoxy powder base coating, and Class 1, nonzinc-filled polyester or epoxy powder base coating, or Class 2, zinc-filled polyester or epoxy base coating. The paint system standards include Mil-C-5541 (Chemical Films and Chemical Film Materials for Aluminum and Aluminum Alloys) and Mil-C-5624 (Turbine, Fuel, Aviation, Grades JP-4 and JP-5). Material vendors are Polymer Corporation, Reading, Pennsylvania, for Type II, Classes 1 and 2, and Ferro Corporation, Cleveland, Ohio, for Type I, Class 1.

In the developmental program, a local vendor is being used to apply the paint. The vendor is using Solids Spray 90XC manual powder coating equipment manufactured by Volstatic, Inc. The equipment provides consistent coating thickness, even on complex surfaces. This portable unit has a 45-lb powder storage drum. The powder is fluidized and delivered through a venturi gun applicator. Total air consumption is minimal, at 6 sft³/min, with good dry powder delivery rates up to 1 lb/min. Constant or variable voltage control provides the electrostatic charge to the powdered particles, which electrostatically bind to the surface being coated. The coated part is fusion-cured in conventional ovens. For this specific application, curing temperatures are between 325°F and 375°F, as compared to the 180°F required by conventional solvent-based paint.

Comparison of electrostatic painting versus conventional painting. Technical and economic advantages of the electrostatic powder painting process over conventional solvent-based painting result in (1) a one-third reduction in curing time, saving both energy and labor, and (2) a reduction in the number of coats per unit from two to one, saving material cost and labor. The material and labor cost savings are estimated to be $1.05/ft² of coated surface. The cost for the electrostatic painting equipment is relatively low; the unit used in the developmental program costs approximately $3,500.

Implementation of the electrostatic powder paint system requires minimal facility change. At the Hughes Tucson plant, the portable powder coating equipment is used in existing conventional wet spray booths. Estimated personnel training time is two weeks. If the use of dry powder painting is expanded, the wet spray booth water scrubber system could be replaced with a dry bag powder overspray collector, which would eliminate the wastewater from these scrubbers. The collected dry residue would still require hazardous waste disposal, but the volume of wastes would probably be less than the wet sludge generated in conventional spray painting because there would be less paint overspray.

Hazardous waste production is minimized using the dry powder painting technology. The coating is dry; therefore, the empty containers are free of residual material and can be disposed of as normal refuse. In addition, because the material is dry, solvent use for cleanup is much reduced, and solvent use for mixing paint formulas is eliminated. The number of paint types needed may also be reduced since powder paint can be used for multiple applications; thus, wastes from partially used containers and stored material with limited shelf-life will be reduced. Using a dry bag overspray collector will minimize the volume of hazardous waste generated by overspray and eliminate the wastewater that would need treatment.

The process modification was successful because it improved both the production rate and quality; it decreased staff requirements and consequently it decreased costs. As in previously discussed case studies, an improvement in production had been the primary motivation for implementing the process modification. The subsequent reduction in hazardous waste generation became a secondary benefit.

Fluidized Bed Powder Coating

Fluidized bed powder coating is typically used to apply relatively thick coatings (10 to 60 mils) to small objects.[7] In the fluidized bed method, a dense cloud is created by passing air through a powder reservoir to create a suspension of powder that behaves like a fluid. The part to be coated is preheated and immersed in the fluidized powder, and the powder fuses to the part. The coated part is then cured in a conventional oven.

Example. At Norfolk Naval Shipyard, Virginia, the repair shop was constantly replacing the heavy-duty springs used for the doors of assault landing craft. While in service, these springs were exposed to salt water and beach sand. The combination of corrosive and abrasive environments was too much for conventional paint coatings. The paint shop foreman jury-rigged a fluidized-bed powder-coating unit out of a 55-gal drum and a blower. The springs were heated in a shop oven, then immersed in a fluidized bed of powdered nylon to coat. The resulting finish was of such a high quality that the foreman never saw powder-coated springs returned to the shop for recoating.

Plasma Spray Powder Coating

Plasma spray powder coating is relatively new and is still mostly in the developmental phase. Dry powder is fed into an extremely hot (5,000°F to 15,000°F) gas stream, where the hot gas melts the plastic and forms a plasma of gas and plastic. The residence time of the powder in the gas is kept very short to prevent material decomposition. The plasma stream is sprayed onto the substrate where a dense, pore-free coating forms as the paint material condenses.

The advantage of this system over other dry powder techniques is that the coating is applied and cured in one step, eliminating the need for subsequent heat treatments. Since the substrate surface temperature does not exceed 185°F, this coating system can be used on substrates that are heat-sensitive. For example, tests have shown that 7075-T78 aluminum alloy was not affected when painted by the plasma technique, but a 10% loss of tensile strength occurred with a curing temperature of 255°F.[8] The plasma technique can also be used for items too large to be cured in conventional ovens. However, personnel protection would be required because of the high temperatures produced in the spray.

WATERBORNE COATINGS

Waterborne coatings are used extensively in industry and on a limited basis by the military. In waterborne coatings (as the term suggests), the carrier is a water solvent rather than an organic solvent.

Fewer hazardous pollutants are generated when using waterborne paints than when using solvent-based paints. The most significant decrease is in VOC emissions, which are almost eliminated in waterborne painting because the volume of solvents used are reduced and the solvents used comply with air pollution control regulations. In addition, no solvents are needed for paint thinning, and the use of solvents for cleanup is greatly reduced. Wastewaters generated from waterborne painting contain fewer toxic organics because of the limited solvents in the paint.

In industry, waterborne paints are normally used where surfaces require only moderate protection and where decorative requirements are most important. Water-bornes are extensively used for decorative or protective coatings on metallic surfaces, as well as for nonmetallic surfaces such as hardboard, wood cabinetry, and plastics.[9]

There are two key disadvantages of waterborne paints. First, the surface must be completely free of oil-type films or the paint will not adhere well. Second, waterborne coatings require longer drying times or even oven drying in cold or humid weather; this requirement may result in significant expense to outfit a facility for waterborne paints.[10]

Positive results with water-based primers have been achieved in applications where conditions were acceptable.

Case Study 6.2: Use of Waterborne Primer on Aircraft

At the Naval Aviation Depot (NADEP) Pensacola, waterborne primers are being tested with the goal of substituting them entirely for existing solvent-based primers. The waterborne primer selected is a water-reducible, amine-cured, epoxy primer manufactured by Deft Chemical Industry, Inc. This waterborne paint does contain some compliance solvent, but less than solvent-based primers. The paint's volatile fraction contains approximately 80% water and 20% solvent. In addition to reducing solvent emissions and wastewater discharge, cleanup is performed effectively using hot water.

Deft water-reducible coatings are supplied in two components—a pigmented resin solution containing corrosion inhibitors and a clean, unpigmented curing agent solution. The two components, which are packaged in a 1-gal kit, must be carefully mixed with 3 gal of deionized or distilled water. The function of the water is solely to control the paint's viscosity during application. After the paint is mixed, the material is catalyzed and is ready for application. The catalyzed coating must be used immediately, since it has a pot life of only 6 hr. After mixing, the primer can be applied using conventional spray paint guns. After the primer is applied, the water evaporates, leaving the nonvolatile coating and some of the unevaporated solvent. At this stage of drying, the film is similar to a high-solids coating without water. The solvent then evaporates and leaves the low molecular weight pigmented material. Reportedly, the final film is physically and chemically identical to an analogous film deposited by a solvent-based coating.

Painters at Pensacola initially tried to spray paint entire H-53 helicopters with water-based primers, relying on solvent-based primers for touchup work only. However, because of the presence of ingrained oil, the primer frequently would not dry properly or adhere adequately to the porous surface, thus making use of water-based primers infeasible for these helicopter surfaces. At first, Pensacola personnel tried to clean helicopter surfaces with freon and alcohol, but the oils remained entrapped on the surfaces. Since approximately 50% of all painting performed at the NADEP is the overspray of whole aircraft, water-based primers have the potential to only partially replace solvent-based primers. Use of the water-based primer on parts, however, proved to be effective.

In the past, approximately 20% of parts painted with solvent-based primers were rejected and had to be repainted. This rejection rate has been reduced to 2% with the new water-reducible primer. However, there are also disadvantages to the waterborne primers, which are slower to dry than solvent-based paints. Some painting supervisors believe that these paints do not provide the same overall corrosion protection, and the paint's inability to adhere to oily surfaces is well documented. Nevertheless, personnel at Pensacola NADEP have found that in most of their applications, water-based primers are superior overall because they are easy to apply, they decrease overspray, they lower the rejection rate, and they make cleanup easier.

Case Study 6.3: Lockheed's Experience with Water-Based Paints

In 1960, the skin paint line was installed to coat aircraft parts to protect them from scratching and corroding during aircraft assembly. From 1960 to 1983, Fabrifilm, a solvent-based coating, was used to protect aircraft surfaces. Water-based coatings were tested in 1983 and are currently being used along with Fabrifilm coatings. However, water-based coatings are used only to provide in-house protection of aircraft surfaces during assembly and are removed prior to final aircraft painting.

Lockheed also tested water-based primers to determine if they could replace solvent-based primers. The company was hesitant to make the change, believing that the useful life of water-based primers was shorter than that of solvent-based primers. Whereas water-based primers meet military specifications for a useful life of 500 hr, solvent-based primers can last up to 2,500 hr. Therefore, Lockheed reportedly does not intend to make the change on a permanent basis, regardless of the quantities of hazardous waste produced, unless the performance of the water-based primer can equal or exceed the performance of solvent-based primers. In addition, Lockheed personnel expressed the belief that solvent-based paints are lighter for the same thickness than water-based paints, less expensive, easier to apply, easier to remove for inspection, and more durable. Solvent-based primers also dry more rapidly than water-based primers. If Lockheed were to make the change, ovens would have to be installed to hasten the drying of painted aircraft parts.

HIGH-SOLIDS COATINGS

High-solids solvent-based coatings, which are similar in composition to conventional solvent-based coatings, are becoming more widely used in industrial applications. These coatings contain about 25% to 50% solids and, compared to solvent-based coatings, use lower molecular weight paint resins with highly reactive sites to aid in coating polymerization. The finished coat is comparable to typical solvent-based coatings.

The general opinion of industry experts is that high-solids solvent-based coatings will become the "standard" to replace regular solvent-based coatings. The major advantage will be the capability to comply with the more stringent VOC limitations while using the same basic paints, equipment, and application techniques.[11]

High-solids coatings require special spray equipment for application because of their high viscosity. Because less solvent is used, less is available to wet metallic surfaces that are contaminated with oils; therefore, surface preparation for removal of oils is more critical. Also, spray application can be wasteful because there is a tendency to apply too much coating to achieve a "wet" appearance similar to that obtained with normal solvent coatings.

Improving the efficiency of paint application reduces waste generation in two ways. Reduced use of paint results in less solvent to evaporate in leaving the final paint film. Reduced overspray means less paint sludge to be removed from a water-wall scrubber, or less paint to be eventually stripped off the walls or floor of a paint booth. Techniques to improve painting efficiency are discussed as follows (and summarized in Table 6.2).

Table 6.2. Expected Transfer Efficiency of Various Painting Methods

Painting Method	Transfer Efficiency
Air-atomized, conventional	30 to 60%
Powder coating	90 to 99%
Air-atomized, electrostatic	65 to 85%
Pressure-atomized, conventional	65 to 70%
Centrifugally-atomized, electrostatic	85 to 95%

Source: "Calculations of Painting Wasteloads Associated with Metal Finishing," U.S. EPA, June 1980.

ELECTROSTATIC PAINTING

Wet electrostatic painting is similar in theory to depositing dry powder coatings by electrostatic attraction (Figures 6.2 and 6.3). It differs in that some solvent is

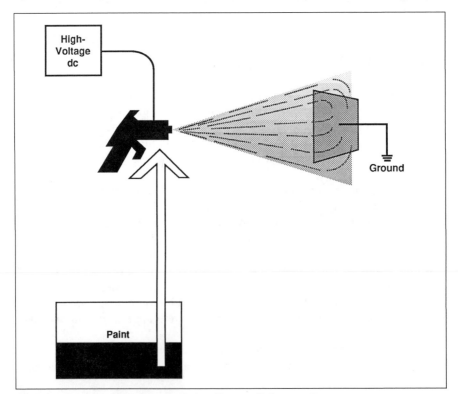

Figure 6.2. Electrostatic spray painting (Source, Reference 3).

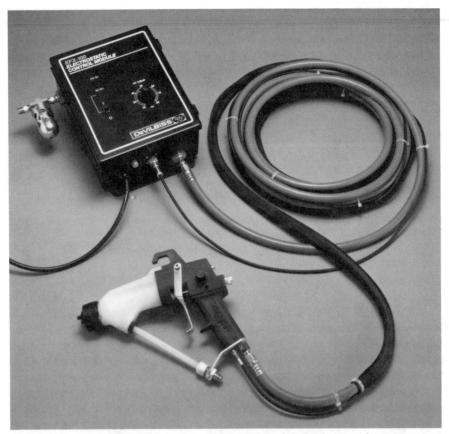

Figure 6.3.   Electrostatic spray paint equipment (Photo courtesy of DeVilbiss Company, Toledo, Ohio).

used as a thinner (the solvent content is lower, however, than in conventional spray painting). Overspray is minimized, if not eliminated, resulting in hazardous waste reduction. Wet electrostatic painting is widely used for painting aircraft parts and other small, complex, nonaluminum metallic articles. There is, however, concern over the potential safety hazard of imparting high voltage to an aircraft that may still contain fuel vapors.

Electrocoating

Electrocoating is similar to metal plating and is commonly used in automotive body coating. In this process, metallic parts or other electrically conductive parts are dipped into a solution that contains specially formulated ionized paint. The action of an electric current induces the paint ions to deposit on the part. The paint formulations are a special class of waterborne nonvolatile organic compounds. Hazardous waste production is minimal, and VOC air emissions are almost eliminated.

One limitation inherent in this process is the requirement for dip tanks. This requirement limits the size of items that can be painted. A more important disadvantage is that the system can be used to apply only one coat (either a prime coat or a single finish coat) because the electrocoated surface prevents further electrodeposition.[8]

Centrifugally Atomized, Electrostatic

A variation of electrostatic painting is the spinning disc or bell type (Figure 6.4). These high-speed discs atomize the paint finer than air-atomization and direct more paint to the target. They are particularly efficient at applying difficult-to-disperse high-solids paints.

Figure 6.4.   Electrostatic spinning disc paint atomizer (Photo courtesy of DeVilbiss Company, Toledo, Ohio).

AIRLESS AND PRESSURE ATOMIZED AIR-ASSISTED SPRAY PAINTING

Airless spray painting can be used for most applications where air spray is used. Airless sprayers have 20% to 30% better transfer efficiency than air spray, resulting in less overspray waste and lower VOC emissions (Figure 6.5). The disadvantages are a coarser finish and a higher paint flow rate, requiring a higher degree of control of paint application.

This system combines the best characteristics of both air and airless spray.[12] It uses an airless fluid spray tip to atomize the coating into a fan pattern at moderate

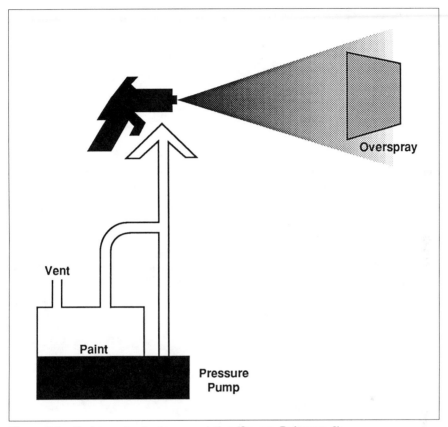

Figure 6.5. Pressure atomized spray painting (Source, Reference 3).

pressures; a second low-pressure air stream is injected just after the nozzle to improve atomization and spray pattern. This new system is reported to provide the finer control of air spraying, with less overspray, and a higher transfer efficiency than airless spray.

Automated Painting

Conveyor systems can be operated to maximize painting efficiency while minimizing the generation of hazardous waste.

Example. In 1980, the Lockheed-Georgia Company installed a modern conveyor system in an automated paint and process line used for painting small aircraft parts at its C-5 assembly plant in Georgia. Painters spray parts as they move along the conveyor system. The conveyor allows parts to be plated, painted twice, and oven-cured, if necessary, all without being touched. The system has improved product quality because impurities from handling are eliminated; in addition, operators can concentrate on improving painting technique, which has reduced overspray and excess paint use. Overall, the system is more efficient and produces less hazardous waste from overspray and cleanup.

In conjunction with the new conveyor system, Lockheed has been retraining operators and inspectors to help them determine the proper paint thickness. The primary reason for the training program is to reduce aircraft weight and paint material cost. If the training program is successful, the quantities of waste solvents and paint sludges should also be substantially reduced.

Robotic control has a tremendous potential for hazardous waste reduction, both in painting and paint stripping applications. Not only can overspray and spills be reduced, but the higher temperatures required for application of ''low'' and ''no'' solvent formulations can be sustained without human discomfort. The use of robotics as currently employed is best suited for industries that mass-produce products or components, such as the automotive industry.

Example. At Lockheed-Georgia, a robotic painting system was installed to reduce paint overspray and improve product quality and efficiency. The robot had the capability to paint an 8-ft by 6-ft rectangular area and could be used for both normal spray painting and electrostatic painting. Usage was discontinued, however, because of difficulty in spraying the irregularly shaped aircraft parts.

REFERENCES

1. Schmitt, G. F., Jr. ''U.S. Air Force Organic Coatings Practices for Aircraft Protection,'' *Metal Finishing* November (1981).
2. Brewer, G. E. F. ''Compliance Solvents for Formulation and Thinning of Spray Paints,'' *Metal Finishing* January (1984).
3. ''Calculations of Painting Wasteloads Associated with Metal Finishing,'' U.S. Environmental Protection Agency, (June 1980).
4. Joseph, R. ''California VOC Rules for the Coating of Metal and Plastic Parts,'' *Metal Finishing* May (1985).
5. Cole, G. E., Jr. ''VOC Emission Reductions and Other Benefits Achieved by Major Powder Coating Operations,'' Paper No. 84-38.1, Air Pollution Control Association (June 25, 1984).
6. Bowden, C. C. ''Powder Coatings on Aluminum Substrates,'' *Metal Finishing* February (1983).
7. Minuti, D. V., and M. J. Devine. ''Innovative Application of Materials for Aircraft Wear Prevention,'' Naval Air Development Center, Warminster, Pennsylvania.
8. ''Electrocoating Today,'' *Products Finishing* February (1983).
9. Albers, R. ''Waterborne Has Solvent Borne Properties,'' *Industrial Finishing* April (1984).
10. Higdon, M. ''Hyster's Experiences: Air Dry, Water-Borne Primers,'' *Products Finishing* April (1981).
11. Nickerson, R. S. ''Applying High-Solids Coatings,'' *Products Finishing* November (1981).
12. Sinclair, R. ''Air-Assisted Airless Spray Painting,'' *Products Finishing* February (1984).

GENERAL REFERENCES

Anderson, C. C. "High Solids Automotive Topcoats," presented at 78th Annual Meeting of Air Pollution Control Association, June 16-21, 1985.

Cole, G. E., Jr. "Powder Coating: 1984," *Products Finishing* January (1984).

Kikendall, T. R. "Converting from Conventional to Compliance Coating Systems," *Industrial Finishing* June (1982).

Schmitt, G. F., Jr. "U.S. Air Force Organic Coatings Practices for Aircraft Protection," *Metal Finishing* November (1981).

Shaffer, P. D. "When to Use Airless Electrostatic Spray," *Products Finishing* February (1984).

CHAPTER 7

Removal of Paint and Coatings

DESCRIPTIONS OF PAINT STRIPPING PROCESSES AND WASTES

Paint stripping is the process of removing paint and paint-type coatings from surfaces, usually as a preparation for inspection, dismantling, repairing, or repainting. In paint stripping, solvents and/or solvent-chemical mixtures are applied to the surface to physically destroy either the paint coating itself or the paint's ability to stick to the surface. When this process is complete, the paint/solvent residue is removed from the surface, usually by pressurized water wash and/or scraping. In many instances, the solvent stripper must be reapplied to remove multiple paint coats and particularly resistant paints.

The wastes generated in the stripping process are a significant source of pollutants. These wastes include the solvent/paint residue, which can be collected separately, and the waste wash water, which contains solids and dissolved chemicals from paints and solvents. Collected solvent/paint residues are normally put in drums and transported to a licensed hazardous waste disposal site. The waste wash water requires treatment in an industrial wastewater treatment plant to remove the paint stripping solvents (usually phenolic or methylene chloride based) and metals picked up from the paint.

Strip baths are also used to remove paint from components. In this method, components are immersed in tanks of stripping solvent. After the solvent dissolves the paint, the stripped parts are removed from the tank and washed with water. The stripping baths are replaced periodically, generally once or twice a year. The hazardous waste solvent/paint liquid and sludge from the bath are then disposed of at a hazardous waste disposal site. The wash water is discharged to an industrial waste treatment plant.

The hazardous and toxic characteristics of the wastes generated at stripping facilities vary considerably. Various paints contain different hazardous constituents (e.g., chromium, cadmium, lead) that affect the degree of hazard and the disposal method to be used. In addition, since paint stripper solvents are formulated from many different compounds, the mixture can also significantly affect waste hazards and disposal methods.

The concentrated waste from stripping baths and surface scraping contains mostly pure solvent and paint residue components with associated hazardous characteristics. However, the degree of toxicity of wastewaters from washing varies based on the type of paint and solvent, the amount of solvent, and the volume of wash water used. Table 7.1 presents typically reported concentration ranges of paint stripping wastewater.[1]

Table 7.1. Paint Stripping Wastewater Characteristics

Parameter	Range
pH, unit	6.2 – 8.0[a]
Phenols, mg/L	17.7 – 45.2
Methylene chloride, mg/L	3.8 – 219.2
Chromium (hexavalent), mg/L	0.10 – 1.12
Total chromium, mg/L	0.164 – 1.187
Cadmium, mg/L	0.024 – 1.09
Lead, mg/L	0.001 – 0.01

[a]Caustic strippers may exceed pH 10.

Another significant source of pollutants is solvent emissions that are discharged into the atmosphere. When solvents are exposed to air, a portion of the solvent is vaporized into the surrounding area. To prevent hazardous working conditions, solvent stripping areas are ventilated with large volumes of fresh air, which remove harmful levels of solvent vapor. The ventilated air is normally discharged to the outside, where solvent vapors are diluted and dispersed.

Air emissions generated in solvent stripping are difficult to quantify. Emissions generally are expected to include the volatile organic compounds (VOCs) found in the solvent itself, mainly methylene chloride and phenols. Little information on the subject is available, however, primarily because these emissions have only recently come under regulation by the EPA and by most state or local agencies. Thus, the need to quantify these emissions to comply with Occupational Safety and Health Administration (OSHA) requirements, which specifically concern on-the-job safety of workers, has been minimal. The EPA is currently developing standards for toxic air emissions, and these limitations will affect solvent stripping operations.

MODIFICATIONS TO CONVENTIONAL SOLVENT PAINT STRIPPING

Several waste reduction techniques have been demonstrated or are practiced by the military and industry.[1] These techniques are generally nontechnical, labor-intensive methods that reduce the volume of hazardous liquids and wash waters.

Example. At Norfolk NADEP, paper is placed on the floor of the paint stripping hangar to collect the loosened paint and the spent stripper solution. This water-free technique has eliminated the high volume of solvent-laden wastewater normally produced in such a facility. The reduced volume of waste is then incinerated.

Example. Reuse of stripping solvent has been investigated at Hill Air Force Base (AFB) as a means to reduce waste generation. Laboratory testing was conducted to test filtering paint particles from collected solvent/paint residues. In theory, the filtered solvent stripper could be reused. Initial tests showed some loss of stripper characteristics, which could probably be overcome by adding makeup chemicals. The major problem of collecting the stripper/paint residue without using either water or significant hand labor was not solved. A full-scale solvent reuse system could save $60,000 per month at Hill AFB if a cost-effective method were found to collect the solvent/paint residue.

Industry has attempted reducing wastes by using labor-intensive methods to collect solvent/paint residue in concentrated form, thereby minimizing the volume of hazardous wastes and wash waters generated.

Example. Pan American Airlines at John F. Kennedy Airport in New York has used aluminum troughs taped to the sides of the aircraft to collect waste solvent/paint residue and to convey it directly into 55-gal drums. This technique minimized wash water use, thus decreasing the waste volume. Plastic troughs and sheets beneath aircraft have also been used to collect stripping waste and to minimize the use of wash water for cleanup. Another practice is using manual squeegees to remove the maximum amount of stripper before washing.

Two other methods of reducing onsite generation of wastes include using paint stripping contractors (who merely move the problem from one place to another) and eliminating use of paint altogether by substituting easily removed decals.

The current practices to reduce waste production in conventional operations for solvent/chemical paint stripping generally have limited benefits, although in certain applications the practices may prove effective. The major problems associated with solvent/chemical strippers have not, for the most part, been solved. Solvent air emissions remain at the same levels because the solvent use is basically the same. Lower volumes of more concentrated solvent wastes are produced, but the total amount is still considerable and is probably even more hazardous to handle because it is more highly concentrated. In addition, wash water is still required for final surface washing, and although the concentrations of contaminants are lower, the wastewater must still be treated to meet local discharge limitations.

ALTERNATIVES TO CONVENTIONAL PAINT STRIPPING TECHNIQUES

Several alternatives to traditional techniques are available for solvent/chemical paint stripping. These alternatives, which reduce the generation of hazardous waste, require new equipment and facilities. The techniques include plastic media blasting (PMB) paint stripping, wet media stripping, laser paint stripping, flashlamp stripping, water-jet stripping, salt-bath paint stripping, burn-off systems, and hot caustic strippers. The more promising developments in the area of paint stripping modifications are summarized below.

PLASTIC MEDIA BLASTING (PMB) PAINT STRIPPING

Conventional sand blasting, abrasive blasting, and glass bead blasting have been extensively used for decades to remove paint and rust from metal surfaces. These removal techniques cannot be used in many applications, however, because the abrasive media can damage aluminum or fiberglass surfaces and small or delicate steel parts. Sand and glass blasting can also cause respiratory ailments, such as silicosis, among workers. Softer dry media (walnut shells, rice hulls, etc.) have been used with limited success for various paint stripping operations where sand and glass could not be used. The "soft media" blasting method has received considerable attention for both military and industrial applications. These natural soft materials are reasonably effective but are difficult, if not impossible, to recycle and are susceptible to bacterial growth during storage, which is reported to cause respiratory infections among users.

Recently, a new medium was developed and manufactured in commercial quantities for blast stripping painted surfaces without damaging the undersurface. The new material has many advantages over other materials, including engineered abrasive characteristics; it is recyclable, durable, and nonhazardous. The material is constructed of soft plastic formed into rough-edged granular media. Old paint is dislodged with conventional sand blasting equipment by using the recoverable plastic media instead of sand, producing a dry waste of pulverized paint and plastic media. The waste volume is significantly reduced, and the waste is more readily disposed of than the wastewater produced in conventional solvent stripping.

United States Plastics and Chemical Company (a former subsidiary of Koppers, Inc.) manufactures the plastic media. The plastic is available in three material hardness grades (Polyextra, Polyplus, and Type 3) and six grain-size sieve distributions (12–16, 16–20, 20–30, 30–40, 40–60, and 60–80). The plastic is used in a wide variety of applications to strip coatings from substrate materials. Conventional blasting equipment can be used, although hoses and nozzles are usually modified to account for the plastic's lower aggression on equipment (Figure 7.1).

Paint stripping by PMB is the most promising alternative to conventional solvent stripping. It has been successfully demonstrated for aircraft renovation at Hill AFB,[2,3] Pensacola NADEP,[4,5] Republic Airlines, United Airlines, and Boeing Vertol. Many other industrial facilities are considering the process because of its highly successful demonstrations and testing as well as its cost-effectiveness.[5-7] Table 7.2 lists suppliers of PMB equipment and supplies.

By carefully controlling the size of the media and the conditions of the process, the plastic media can be separated from the loosened paint particles and recycled. PMB completely eliminates the generation of wet hazardous waste (solvents and paint

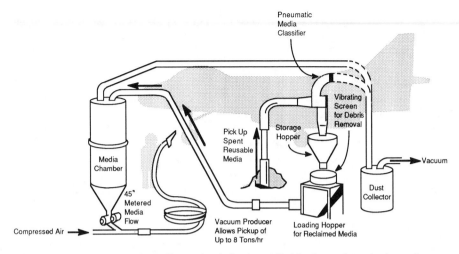

Figure 7.1. Schematic of a self-contained plastic media blasting paint stripping unit.

Table 7.2. Suppliers of Plastic Media Blast Paint Stripping Equipment

Company	Address	Phone
Aerolyte (Clemco)	1657 Rollins Road Burlingame, CA 94010	(415) 570-6000
Caber, Inc.	745 South Chicago Seattle, WA 98108	(206) 762-1320
Pauli & Griffin	907 Cotting Lane Vacaville, CA 95688	(707) 447-7000
Schmidt Manufacturing	11927 South Highway 6 Fresno, TX 77454	(713) 431-0581
U.S. Technology	328 Kennedy Drive Putnam, CT 06260	(203) 928-2707
Zero Manufacturing Co.	811 Duncan Avenue Washington, MO 63090	(314) 239-6721

sludge in water); however, PMB does produce a small volume of dry waste, which is classified as hazardous because of metal content.

The two key parameters for successful use of PMB are hardness and reusability. First, the plastic media must be harder than the paint, but softer than the surface underneath the paint coat. Second, the media must be durable enough to be reused repeatedly to minimize the amount of residue that must be disposed of as a hazardous waste.

With some very hard paints (such as epoxy and urethane paints), presoftening with a solvent (such as methylene chloride) has been used before PMB stripping. However, recent test data provided by the media supplier indicate that, with modifications to media selection and application methods, these paints can be successfully removed without presoftening the paints.

The PMB technique has been effective in stripping and removing a variety of coatings from a number of substrate surfaces. However, extreme care must be exercised on composite surfaces, thin-skinned aluminum, and other fragile materials. In particular, composite fibers have sometimes unravelled when composite surfaces that did not have a resin-rich surface were blasted. In some instances, using excessive pressure or holding the nozzle too close has resulted in surface damage. Even though the PMB process is relatively simple, considerations such as these make it imperative that operators receive adequate training.

The blasting action of the PMB technique helps to stress-relieve surfaces when removing paint from titanium, stainless steel, alclad, and anodized aluminum. Alclad aluminum surfaces have a sandblasted appearance after blasting because the aluminum cladding is softer than the plastic media. This soft aluminum coating is shifted, but not removed; in fact, after blasting it presents a much better surface for repainting.

Many additional applications will be developed as testing continues. Table 7.3 lists some of the coatings and substrates successfully stripped with plastic media.[8]

Table 7.3. Applications for Paint Stripping by PMB

Coatings	Substrates	Applications
Polyurethane	Aluminum	Aircraft fuselage
Epoxy polyamide	Alclad aluminum	Components
Acrylic lacquer	Anodized aluminum	Ship bilges
Enamel	Steel	Vehicle bodies
Fluorocarbons	Magnesium	Boat hulls
Metallic spray	Anodized magnesium	Engine components
Koropon primer	Titanium	Truck wheels
Rain erosion	Carbon graphite	Propeller blades
Fuel sealants	Fiberglass (except	Molds
Structural adhesive	Radomes and Kevlar)	Heat exchangers
Corrosion buildup	Honeycomb	Alloy fuel tanks
Lubricants		
Polysulfide sealants		
Carbon buildup		

The potential savings associated with PMB stripping are substantial. The estimated savings in labor, chemicals, and waste treatment or disposal amount to more than $100 million annually.[6] In addition, savings resulting from decreased energy costs and costs for compliance with future environmental regulations coupled with increased productivity are likely to be equally significant. The preliminary cost estimate presented in Table 7.4 illustrates the potential savings.

The plastic media paint removal process is so simple and efficient that it lends itself to a wide variety of uses. The most notable feature of the process, however,

Table 7.4.   Estimated Savings from Adopting PMB at DOD Facilities

Item	Solvent/Chemical Stripping	Plastic Media Stripping
Labor and material		
Work-hours	3,360,000 hr	1,426,000 hr
Solvents/chemicals	7,000,000 gal	0
Wash water	100,000,000 gal	0
Wastes	107,000,000 gal	500,000 lb dry
Operating costs		
Work-hours	$136,516,800	$ 67,698,380
Material supplies	30,960,000	4,400,000
Waste treatment and disposal	8,000,000	1,500,000
Total operating costs	$175,476,800	$ 73,598,380
ANNUAL COST SAVINGS		$101,878,420

is its near-elimination of pollution and toxic waste. The only waste from this system is a comparatively small amount of dry fine plastic dust and paint particles that contain trace heavy metals from the paint. This waste is easily contained within sealed drums and can be safely transported for disposal or storage. No liquid waste is generated. Because the air system can be self-contained and dust removal facilities provided, air pollution can be eliminated. The plastic media are recycled and reused with little degradation. Energy, materials, labor, and product efficiency all cost significantly less than in conventional operations for solvent paint stripping. However, to protect workers from occupational hazards (dust, noise), hearing protection, goggles or masks, and filtered breathing air must be provided.

The following case study illustrates the specific application of PMB technology at Hill AFB.

Case Study 7.1: Development of Plastic Media Blasting (PMB) Paint Stripping at Hill Air Force Base

Hill AFB in Ogden, Utah, has been the lead facility in developing and testing plastic media blasting. The development of this process modification demonstrates the key elements necessary for successfully implementing a waste minimization project. The process itself is elegant in its simplicity—conventional sand blasting equipment was adapted to include media recovery and separation from the waste paint chips and dust.

Initially, Bob Roberts, the Champion of the process, was motivated to replace the existing wet-solvent process, which was environmentally objectionable and occupationally hazardous. Following extensive testing on aircraft components to demonstrate the newer process's effectiveness and safety, personnel at Hill AFB completely stripped an F-4 fighter plane in July 1984. This test demonstrated that the process was much less labor-intensive and was occupationally less hazardous than solvent stripping. The aircraft was completely stripped in 40 work-hours versus 340 work-hours required for wet paint stripping. In addition, greater control in stripping was achieved, compared with wet paint stripping and sanding. This increased control resulted in reduced damage to underlying surfaces.

A full-sized hangar for plastic bead blasting (Figure 7.2) was constructed because

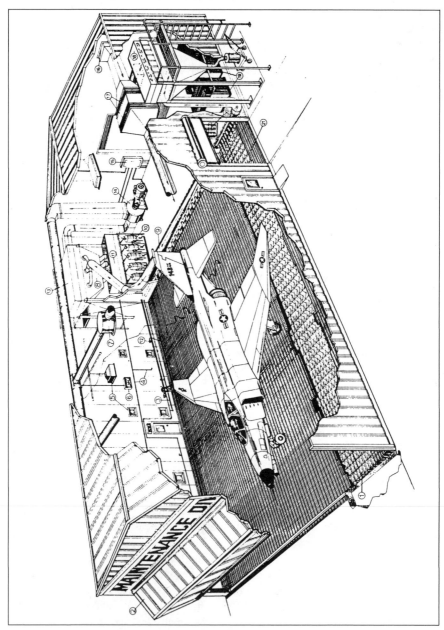

Figure 7.2. Plastic media paint stripping hangar, Hill Air Force Base (Drawing courtesy of Royce Mechanical Systems, Ogden, Utah).

of the promise of reduced personnel requirements and a favorable environmental impact. The hangar incorporates both a live floor-vacuum system to provide ventilation and dust removal and a separation system for bead recovery and reuse. The hangar was funded under a Productivity Enhancement Capital Investment fund, allowing the demonstration facility to be built within one year rather than being built according to the standard military construction schedule. Photographs of the facility are shown as Figures 7.3 through 7.12.

Figure 7.3.   Plastic media blasting paint stripping facility at Hill Air Force Base.

Figure 7.4.   Hill Air Force Base plastic media blasting paint stripping equipment.

Figure 7.5.   Plastic media blasting paint stripping of F-4 aircraft (floor level view).

Figure 7.6.   Plastic media blasting paint stripping of F-4 aircraft (view from above).

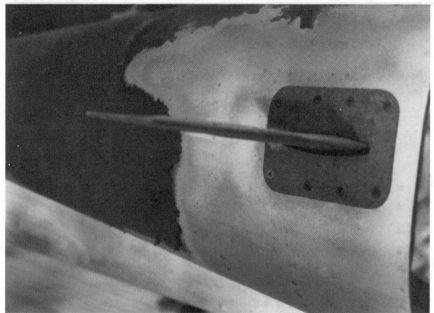

Figure 7.7.   Plastic media blasting paint stripping of anodized aluminum surface.

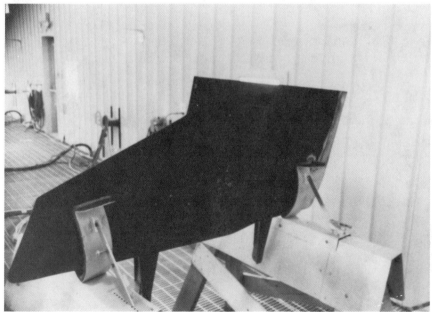

Figure 7.8.   Plastic media blasting paint stripping of F-4 aircraft component.

Figure 7.9. F-4 aircraft component after stripping.

Figure 7.10. Plastic media blasting paint stripped aircraft (front view).

Figure 7.11. Plastic media blasting paint stripped aircraft (view from above).

Figure 7.12. F-4 aircraft stripped by conventional solvent.

Based on initial prototype testing at Hill AFB, estimated savings from using plastic media compared with using solvent paint stripping are summarized in Table 7.5. The quantity and cost savings estimates are based on stripping 215 F-4 aircraft annually. Bob Roberts of Hill AFB prepared these estimates, based on the following assumptions.

Table 7.5. Estimated Savings from Adopting PMB at Hill AFB

Item	Potential Savings	Annual Cost Savings
Hazardous waste	Generates 1/100 the waste sludge, which requires hazardous waste disposal	$ 218,000
Wastewater pollution	Eliminates generation of 210,000 gal/day of wastewater, which must be treated in on-base waste treatment plant before discharge to the city municipal treatment plant	$ 526,375
Materials	Eliminates the use of chemical solvents and requires minimal use of plastic media to make up for worn-out media	$1,091,340
Labor	Requires 1/10 the labor	$2,179,060
Energy	Requires 1/10 the energy	$ 223,929
Flow days	Provides increased flow-day utilization of aircraft	$1,353,210
TOTAL ANNUAL SAVINGS FOR 215 F–4 AIRCRAFT		$5,591,914

Hazardous Waste

The AFB's existing wastewater treatment plant produces approximately 3,000 tons per year of 10% solid weight sludge, which is hazardous. The sludge is transported by truck to California, where it is disposed of at a licensed hazardous waste disposal site for a total cost of $200/ton.

Containing solvent and paint residue, the wastewater generated from solvent stripping F-4 aircraft is estimated to contribute 35% of the total sludge produced from Hill AFB. Therefore, total sludge contributed by solvent stripping is 1,050 tons. The only hazardous waste produced by plastic media stripping is the dry stripped paint residue, which amounts to 120 lb per aircraft, and the dry spent plastic media, which amounts to 200 lb per aircraft—a total of only 34 tons of hazardous waste per year. Thus, the annual reduction in hazardous waste products is 1,016 tons, a 97% reduction, and $218,000 in costs.

Wastewater Pollution

Thirty-five percent (210,000 gal/day) of the 600,000 gal/day of wastewater treated in the on-base industrial waste plant is generated by the solvent stripping operations. Approximately 20,000 to 30,000 gal of water are used to wash off the stripper and paint residue for each stripper application. Several applications of stripper are normally required. Water is also used to wash floors and to provide general area maintenance, further contributing to the wastewater flow.

The annual cost of treatment chemicals at the industrial waste plant is $912,500. Reducing the waste flow by 35% is estimated to reduce treatment chemical use proportionally, thereby saving $319,375 annually. Additional savings in operation and maintenance expenses for the industrial waste plant (i.e., labor and equipment repair and replacement) were estimated at $207,000 annually. Thus, the total estimated annual chemical and operation and maintenance cost savings are $526,375.

Materials

Solvent stripping an F-4 aircraft requires 468 gal of chemical stripper at a cost of $11.40/gal and 12 rolls of aluminum masking tape at a cost of $7.30/roll. The total cost per aircraft is $5,422, or $1,165,902 annually. Plastic media is recycled, but losses occur as a result of abrasion. Media loss is estimated to be 200 lb per aircraft, and at a cost of $1.73/lb, the cost per aircraft is $346, or $74,390 annually. Savings in material costs per aircraft amount to $5,076, or $1,091,340 in annual savings.

Labor

One of the most significant advantages that media stripping has over solvent stripping is its lower labor costs. Solvent stripping an F-4 aircraft requires 341 work-hours per aircraft, at a labor rate of $33.56/hr, for a cost per aircraft of $11,444 ($2,460,460 annual cost). Plastic media stripping is estimated to require 39 work-hours per aircraft, at the same $33.56/hr labor rate, for a cost per aircraft of $1,309 ($281,400 annual cost). Estimated annual labor savings amount to approximately $2,179,060. Typical prototype production comparisons for F-4 aircraft components and estimated production comparisons for other aircraft and equipment are shown in Table 7.6.

Energy

Two components make up most of the energy use in stripping: the energy required to maintain the building interior at the required temperature, and the energy required to operate the equipment's electrical motors. Solvent stripping operations require

Table 7.6.   Production Comparisons of PMB vs Solvent Stripping at Hill AFB

Item	Solvent Strip Time	Plastic Media Strip Time
F-4 Component:		
Rudder	3 hr 36 min	15.6 min
Inboard leading edge flap	2 hr 48 min	21.6 min
Spoiler	40 min	14.4 min
Outboard leading edge flap	2 hr 48 min	18.6 min
Aileron	6 hr 28 min	32.4 min
Wingfold	8 hr 45 min	54.1 min
Stabilator	9 hr 49 min	55.2 min
Aircraft and equipment:		
F-4 (prototype)	341 hr	39 hr
F-100 (museum aircraft)	290 hr	25 hr
P-8 pumper (fire truck)	52 hr (sanding)	4 hr
D-50 Pickup (compact)	40 hr (sanding)	1 hr 20 min
1/2 Ton Pickup (full size)	60 hr (sanding)	1 hr 55 min

significant energy to heat the building because the building interior air must be maintained at $72°F \pm 2°F$ for proper solvent action on the paint, and large fresh air flows are required to ventilate the solvent vapor emissions. These annual heating costs are $201,600 (basis: 507,000 ft³/min fresh air, average annual temperature 51°F, 16 hr/day at 260 days/year building use, steam cost $5.59/million Btus). Solvent stripping mechanical equipment also requires significant electrical energy to operate.

The annual electrical energy cost is $49,634 (Basis: 320-hp motor, 16 hr/day at 260 days/year equipment use, $.05/kWh electrical energy cost). Plastic media stripping operations require much less electrical energy for heating and equipment. Almost no building heating is required because the waste heat generated from the air compressor equipment used in blasting plastic media generates sufficient heat to warm the building. And, because no solvents are emitted into the air (which would have required large fresh air flows), no significant amount of energy is required to heat fresh air. Electrical energy requirements for plastic media stripping equipment are also much lower than those for solvent stripping equipment. The plastic media equipment requires only $27,305 in annual electric energy costs (basis: 340-hp motor, 8.25 hr/day for 260 days/year, $0.051/kWh electrical energy cost). Therefore, plastic media stripping operations are projected to save $223,929 in annual energy costs.

Flow Days

Plastic media stripping decreases the overall time needed to renovate aircraft for use; therefore, aircraft utilization is increased. Plastic media stripping requires 1/2 day, compared to 3-1/2 days to complete the solvent stripping process. Based on USAF cost and planning factors, AFR 173-3, the flow day efficiency cost savings amount to $1,353,210 annually.

The new plastic media stripping facility (also called "blast booth") at Hill AFB

Building 223 is a full-scale plastic media aircraft stripping facility specifically constructed for F-4 aircraft maintenance. The major components of the plastic media paint stripping facility are shown in Figure 7.2. Bob Roberts, program manager, spearheaded the facility's design and construction. Royce Mechanical Systems, Ogden, Utah, provided fast-track component design and facility construction. The facility includes a steel prefabricated insulated building (45 ft × 75 ft × 25 ft high) and all process and support mechanical and electrical equipment. Total facility cost, including equipment and labor, was $647,389. The facility cost payback will be just over one month, based on operation cost savings.

LASER PAINT STRIPPING

In the Air Force's testing of lasers for removing paint, research has been directed at the development of a pulsed carbon dioxide (CO_2) laser system. The pulsed laser was chosen to minimize energy consumption; CO_2 was selected because its wavelength is readily absorbed by paint. Actual pilot-scale tests showed that the paint material was completely removed from test surfaces. The system, in which a full-scale operational installation would be based on a robot-operated pulsed CO_2 laser, is still in the experimental stage, however.

Although this alternative appears to be technically feasible, there are many unknown factors involving system reliability, effects on aircraft substrate and components (electronics, sensors, etc.), air pollutants, and other factors that need extensive research and development work. It may take 10 years or more for this technology to be commercially available. In addition, by one estimate, the initial capital outlay for an aircraft-size facility with automated laser system would be at least $10 million, which is an order of magnitude greater than a comparable plastic media blasting facility.

FLASHLAMP STRIPPING

Flashlamp stripping is similar to stripping with laser light, but it uses high-energy quartz lamps to vaporize paint. The Air Force is conducting research and development (R&D) on this process. Unlike laser stripping, flashlamp stripping has been proven not to harm aircraft electronics. However, this technique is difficult to use and requires extensive operator training. Unresolved questions involve potential damage to various substrates, generation of toxic air pollutants, and design issues regarding a production unit. In Navy tests, this method failed to remove barnacles from the bottom of ships, and the equipment produced loud, annoying "bangs" when operating.

DRY ICE BLASTING

The Lockheed Company investigated dry ice blasting for removing aircraft paint.[9]

Dry ice or carbon dioxide pellets were used as a blasting media. The attractive aspect of this technology is that dry ice pellets vaporize after being used and the only waste product is the dry paint chips. There are, however, questions concerning the potential damage to surfaces, effectiveness of paint removal, and operation costs. One problem is that the generation of carbon dioxide displaces oxygen in a room, necessitating the use of a contained air supply when blasting. Production of fog from humid air is also a problem.

CRYOGENIC COATING REMOVAL

This method operates on the principle that organic coatings become brittle and tend to debond from substrate metals at low temperature because of differential thermal contraction of the coating and the substrate metals. Small cabinet-size equipment based on cryogenics is commercially available. Liquid nitrogen is sprayed on the coating to lower the surface temperature to $-100°F$, and plastic media is mechanically thrown at the surface to break off the frozen paint.[10] This system is not suitable for large-scale operations.

HIGH-PRESSURE WATER-JET BLASTING

Both the Air Force and the Navy investigated water-jet blasting for removing paint. To remove paint, this process uses pulsed or continuous water-jet blasting produced by high-pressure pumping. As with the other systems discussed, the use of a water-jet is technically feasible; however, questions need to be resolved about the system's control and reliability, potential damage to surfaces, ability to remove a wide range of coatings, and worker safety. High-pressure water-jet blasting is currently used by the automotive industry to remove paint buildup from the floor gratings of paint booths.

SALT-BATH PAINT STRIPPING

Equipment is commercially available to strip paints in molten salt baths operating at a temperature of $900°F$.[10] This method is practiced in the automotive and appliance manufacturing industries. In this process, items to be stripped (generally steel) are immersed in the molten salt bath (mixture of sodium hydroxide, sodium or potassium nitrate, sodium chloride, and catalysts) in which the heat destroys the paint. This process cannot be used on parts or equipment constructed of aluminum, nonmetallics, and alloys, because of effects of heat on these materials.

BURN-OFF SYSTEMS

High-temperature flames, ovens, and fluidized beds are commercially used to literally burn paint off. This technology is limited to steel parts.[10]

HOT CAUSTIC STRIPPERS

Hot caustic solution stripping is commercially practiced by industry, and equipment is readily available. Hot caustic baths, typically at temperatures over 200°F, are very effective in removing caustic-sensitive paints. Applications are limited, however, because many of the coatings, such as epoxies, are both caustic and heat-resistant. Stripping is also limited to steel parts because the caustic corrodes many materials, including aluminum.[10]

REFERENCES

1. Law, A. L., and N. J. Olah. "Initiation Decision Report: Aircraft Paint Stripping Waste Treatment System," Naval Facilities Engineering Command, Technical Memorandum, TM No. 71-85-06, Naval Civil Engineering Laboratory, Port Hueneme, CA (December 1984).
2. Roberts, R. A. "Plastic Bead Blast Paint Removal," Aircraft Division, Directorate of Maintenance, Hill Air Force Base, Ogden, UT (March 1, 1985).
3. Roberts, R. A. "Mechanical Paint Removal System: Special Report on Plastic Impact Cleaning Media," Hill Air Force Base, Ogden, UT (July 31, 1984).
4. "Preliminary Economic Analysis of Paint Stripping Using Plastic Impact Media at NARF Pensacola," NARF Pensacola, FL (March 28, 1984).
5. "Coating Removal via Plastic Media Blasting," Materials Engineering Division, NAVAIR, Engineering Support Office, Naval Air Rework Facility, Pensacola, FL (July 18, 1984).
6. Boubel, R. W. "Evaluation of Dry vs. Wet Paint Stripping," Memorandum for the Record to Peter S. Daley, Office of the Secretary of Defense (August 1, 1984).
7. "Status Report on Plastic Media Paint Stripping (F-45 Paint Strip)," Engineering Report No. 002-85. NAVAIR WORK FAC NESO, North Island, San Diego, CA (March 1985).
8. "Synopsis of Testing Performed by U.S. Military Facilities Evaluating U.S. Plastic and Chemical Corporation Plastic Abrasives," U. S. Plastic and Chemical Corporation (September 30, 1984).
9. Schmitt, G. F., Jr. "U.S. Air Force Organic Coatings Practices for Aircraft Protection," Metal Finishing (November 1981).
10. Mazia, J. "Paint Removal (Stripping Organic Coatings)," Metal Finishing Guidebook Directory (1985).

GENERAL REFERENCES

Whinney, C. "Blast Finishing—Part 1." *Metal Finishing* (November 1983).

Whinney, C. "Blast Finishing—Part 2." *Metal Finishing* (December 1983).

Whinney, C. "Blast Finishing." *Metal Finishing* (January 1985).

CHAPTER 8

Waste Treatment to Minimize Disposal

In considering a waste minimization project, the first goal is to try to eliminate the waste as early in the production process as possible. It is, however, rarely feasible to optimize a production process to the extent that no waste is produced. "Zero discharge" is a laudable goal, but it is seldom practical. However, even when the volume of waste from a production process is reduced to the greatest extent possible, waste minimization opportunities do not end. The waste can still be processed to recover useful materials or treated to reduce its volume, toxicity, or mobility in the environment. This final chapter discusses and illustrates with the following case studies a few of the hazardous waste treatment technologies employed to reduce the costs of waste disposal.

Topics covered include:

- recovery of a useful product
- segregated treatment to reduce hazardous mixtures
- toxicity reduction
- volume reduction
- incineration to reduce volume, toxicity, and mobility

RECOVERY OF A USEFUL PRODUCT

Conventional treatment methods for nondestructible inorganic wastes rarely do more than concentrate the harmful constituents in as small a volume as is feasible for disposal. Metals cannot be destroyed but are merely transformed from a dissolved state to a solid form (usually as a hydroxide). One way of eliminating the hazards associated with the disposal of nondestructible waste constituents is by separating and recovering their constituents as useful products.

Three case studies are presented that illustrate the recovery of a useful product from a waste treatment process. The first involves recovery of copper metal from a printed circuit board manufacturing facility using ion exchange and electrowinning. The second case study describes the recovery of vanadium from the wastewater

153

of an oil-fired electric power generating station. The third example concerns a wastewater recycling system for an automobile manufacturing facility. These three examples illustrate that useful products can be obtained from an end-of-pipe treatment facility, and that these recovery processes can reduce the costs and liabilities associated with conventional treatment and disposal.

Case Study 8.1: Recovery of Copper at a Printed Circuit Board Facility

Background and Objectives.   An aerospace manufacturer operates a prototype circuit board fabrication facility in New England. CH2M HILL was selected to design a waste treatment plant for this facility. The principal goal of the design was to minimize the volume of sludges generated by operation of the treatment plant.

Effluent from the treatment plant discharges to a publicly owned treatment works (POTW), so the effluent quality has to comply with federal and local pretreatment discharge limits. Wastes to be treated in the facility consist mainly of continuous overflow rinses from scrubbing, cleaning, electroless plating of copper, electroplating of copper, photoresist processes, and etching (Table 8.1). This mixed waste is characterized as containing a mixture of regulated metals (principally copper and lead) and complexing agents from electroless copper plating and etching processes.

Table 8.1.   Continuous Rinse Flows at a Printed Circuit Board Fabrication Facility

Existing Rinse Flow (gal/min)	Associated Process Tank
4	Acid copper sulfate plating (H_2SO_4)
6	Cleaner conditioner (alkaline)
10	Sodium persulfate etch (H_2SO_4)
6	10% H_2SO_4
6	Catalyst (HCl, Pb, Sn)
6	Accelerator (HBF_4)
6	Electroless copper plating
2	25% HCl
5	Rapid Fixer (H_2SO_4)
5	Photoflow
8	Photoresist developer
4	Cupric chloride etch
4	Ammonium etch (alkaline)
15	Scrubber
87	Total existing flow

Current maximum flow is approximately 90 gal/min. The treatment facility is designed for 50 gal/min with a hydraulic capacity of 90 gal/min. Flow reduction and diversion of noncontact cooling are expected to reduce flows to a range of 15 to 30 gal/min. The presence of complexing agents would render conventional hydroxide precipitation treatment ineffective. One effective treatment for this type of waste has been treatment with large doses of ferrous iron, which displaces copper and lead from their complexes, precipitating metal hydroxides. This process, however, produces voluminous sludge because of the precipitation of iron as well as copper and other metal hydroxides.

Process Description.   To minimize hazardous waste, CH2M HILL recommended treating the combined rinsewater waste by chelating ion exchange. Chelating ion exchange resins, when operated at a pH of 4 to 5, are sufficiently selective for copper that the metal can be removed from the complexing agents, allowing the complexing agents to pass through to the effluent.

Regeneration of the ion exchange resin with sulfuric acid produces a mixture of sulfuric acid and copper sulfate, which is then electrowinned (electroplated) onto flat stainless steel plates, producing sheets of elemental copper that can be peeled off and sold as scrap. This process contrasts with conventional metal hydroxide precipitation, which produces a voluminous sludge requiring disposal as a listed hazardous waste.

In addition to the ion exchange regenerant solution, concentrated copper-bearing wastes (electroless copper plating "growth" and persulfate etch) may be more efficiently electrowinned directly, rather than first being passed through ion exchange.

A process flow diagram for the treatment system is shown as Figure 8.1. Design criteria and equipment sizing are listed in Table 8.2.

Rinsewaters are pumped directly to a mixed pH adjustment tank. Concentrated acidic and caustic wastes are pumped to holding tanks to be metered into the mixing tank to prevent slug loading of acids, alkalis, or metals to the treatment system. In the pH adjustment tank, the waste is adjusted to a pH between 4 and 5 using sulfuric acid or caustic soda (sodium hydroxide), since this pH range is optimal for selective ion exchange treatment for copper.

The waste is then pumped through duplex cartridge filters. Cartridge filters were selected because of the low suspended solids in the wastewater. It is estimated that cartridge replacement will be required once per week. The effluent from the filter passes through two activated carbon vessels in series. Activated carbon adsorption is provided to remove organics that could foul the ion exchange resins, reducing the resins' capacity and resulting in more frequent replacement. Two carbon adsorption columns are provided in series, thus allowing complete use of the carbon in the first vessel, with the second providing polishing. Following complete breakthrough of organics from the first carbon column, that carbon should be replaced and the roles of the two vessels reversed. It is estimated that carbon replacement will be required annually. The carbon vessels will be backwashable, with the backwash water to be returned to the initial pH adjustment tank. At this facility, because of the characteristics of the organics present, the use of carbon is considered to be optional. There is a cost tradeoff between increased ion exchange resin usage and the capital and operating cost of the carbon system.

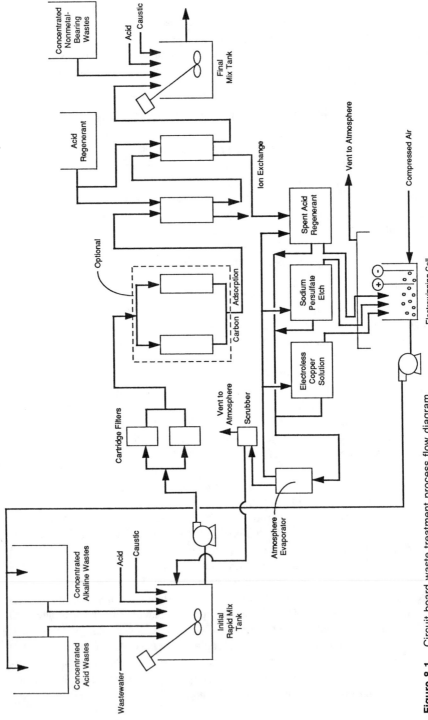

Figure 8.1. Circuit board waste treatment process flow diagram.

Table 8.2. Circuit Board Treatment Facility Design Criteria and Equipment Sizing

Process Parameter (units)	Criteria	No.	Description
Acid holding tank			
Tank volume (gal)	1,500	1	FRP, vinyl lined
Transfer pumps (gal/min)	1	2	Diaphragm metering, 316 SS, Teflon diaphragm
Alkali holding tank			
Tank volume (gal)	1,500	1	FRP, vinyl lined
Transfer pumps (gal/min)	1	2	Diaphragm metering, 316 SS, Teflon diaphragm
Initial pH adjustment			
Design flow (gal/min)	50		
Reaction time (min)	20		
Reaction volume (gal)	1,000		
IX, Carbon backwash (gal)	400		
IX Rinse (gal)	600		
Tank volume (gal)	2,000	1	FRP, vinyl lined
Mixer (HP)	1	1	Vertical shaft, 316 SS wetted parts
Process pumps (gal/min)	45	3	Centrifugal, ductile iron, 316 SS impeller
Duplex cartridge filters		1	
Design flow (gal/min)	50		
Activated carbon columns			
Design flow (gal/min)	50		
Empty bed contact (min)	5		
Carbon volume (ft^3)	33		
Vessel diameter (ft)	2		
Cross-sectional area (ft^2)	3		
Backwash rate ($gal/min/ft^2$)	15		
Backwash rate (gal/min)	45		
Backwash time (min)	10		
Backwash volume (gal)	400		
Ion exchange columns		2	
Design flow (gal/min)	50		
Loading rate ($gal/min/ft^3$)	2		
Resin volume (ft^3)	25		
Vessel diameter (ft)	3		
Cross-sectional area (ft^2)	7		
Backwash rate ($gal/min/ft^2$)	4		
Backwash rate (gal/min)	30		
Backwash time (min)	10		

Table 8.2.   Continued

Process Parameter (units)	Criteria	No.	Description
Backwash volume (gal)	300		
Exchange capacity (meq/mL)	0.5		
Regenerant (bed volumes)	4		
Acid concentration	5%		Sulfuric acid
Final pH neutralization			
Design flow (gal/min)	50		
Reaction time (min)	20		
Reaction volume (gal)	1,000		
Surge storage (gal)	500		
Tank volume (gal)	1,500	1	FRP, vinyl lined
Mixer (hp)	1	1	Vertical shaft, 316 SS, wetted parts
IX Regenerant holding tank			
Tank volume (gal)	1,500	1	FRP, vinyl lined
Transfer pumps (gal/min)	1	2	Diaphragm metering, 316 SS, Teflon diaphragm
Persulfate etch holding tank			
Tank volume (gal)	720	1	FRP, vinyl lined
Transfer pumps (gal/min)	1	1	Diaphragm metering, 316 SS, Teflon diaphragm
Electroless copper growth holding tank			
Tank volume (gal)	460	1	FRP, vinyl lined
Transfer pumps (gal/min)	1	1	Diaphragm metering, 316 SS, Teflon diaphragm
LSA Electrowinning system			
Copper loading (gal/wk)	54,000		
Electrowin rate ($gal^3/ft^2/hr$)	15		
Surface area required (ft^2)	75		
Power ($amps/ft^2$)			
Recirculation pumps (gal/min)	5	2	Centrifugal, ductile iron, 316 SS impeller
Effluent copper (mg/L)	1,000		
Atmospheric evaporator		1	
Evaporation rate (gal/hr)	2–5		
Blower capacity (sft^3/min)	1		
Heat (Btu/gal)	8,090		
Recirculation pump (hp)	1		
Evaporation rate (gal/mo_x)	1,500		

(cont.)

Table 8.2.   Continued

Process Parameter (units)	Criteria	No.	Description
HMT Electrowinning system			
Copper loading (gal/wk)	54,000		
Electrowin rate (gal^3/ft^2/hr)	6–39		
Power			
Power (kWh/lb^3)	4		
Recirculation pumps (gal/min)	5	2	Centrifugal, ductile iron, 316 SS impeller
Effluent copper (mg/L)	15		

The wastewater then flows through dual ion exchange columns, operating in series, utilizing a chelating cationic ion exchange resin. Provisions are made to backwash the resin with city water. Following exhaustion, the lead ion exchange resin is regenerated with sulfuric acid, rinsed with city water, and returned to service as the lag, or polishing, unit. The backwash and rinsewaters are returned to the rapid mix tank for treatment. The waste regenerant is piped to a holding tank for copper recovery by electrowinning.

The effluent from the ion exchange units discharges to the final rapid mix tank, where the pH is adjusted with sulfuric acid or caustic soda and discharged to the city sewer.

The spent ion exchange regenerant, electroless copper plating bath growth, and sodium persulfate etching solutions are stored in separate holding tanks. These solutions are pumped to a low surface area (LSA) parallel slate type, electrowinning unit for recovery of metal. The electrowinning unit is batch operated. When a batch is treated, it is pumped to the acid or alkaline holding tank and bled into the influent pH adjustment tank for retreatment. The effluent from the LSA electrowinning cell is expected to contain 1 or 2 grams of copper per liter.

A high surface area type, high mass transfer (HMT) electrowinning system is being considered for use instead of an LSA system. The advantage of an HMT system is its reported ability to reduce effluent copper concentrations to a few milligrams per liter. This equipment would reduce the load of copper on the ion exchange system and thus result in less frequent regeneration. Disadvantages of the HMT are increased power consumption and lower rates of metal recovery (resulting in longer electrowinning times). An HMT system consists of carbon fiber cathodes that provide a high surface area for plating. When saturated with metal, the carbon electrodes are transferred to an electrorefining cell LSA (a modified electrowinning unit), where metal is anodically stripped off the carbon electrode and deposited on flat stainless steel electrodes for metal recovery.

As an alternative to HMT electrowinning, an atmospheric evaporator, which would increase the concentration of metal in the ion exchange regenerant prior to electrowinning, is also being considered. Benefits include volume reduction, improved electrowinning efficiency due to increased metal concentration resulting in reduced frequency of ion exchange regeneration, and potential reuse of ion exchange regeneration acid following dilution to the original acid concentration. This reduction in volume is beneficial to operation of the system in that fewer batches are required to be electrowinned. The required electrowinning cathode area is unchanged, and the time for electrowinning is unchanged since these are dependent on the mass of metal to be removed rather than hydraulic volume throughput. Disadvantages of evaporation include the requirement for heating an acidic solution, safety problems associated with handling a heated concentrated acid, and additional power and maintenance requirements.

Cost Estimates.   Two systems are considered for installation at this facility: LSA electrowinning with an atmospheric evaporator and HMT electrowinning. Order-of-magnitude cost estimates for the installation of the two systems are provided as Table 8.3. Operating costs are listed on Table 8.4.

Table 8.3. Order-of-Magnitude Capital Cost Estimate
Circuit Board Treatment Facility

Category Items		LSA Elect. Alternative	HMT Elect. Alternative
EQUIPMENT			
Tanks		$ 50,200	$ 50,200
Mixers		13,200	13,200
Pumps		47,300	47,300
Duplex cartridge filters		10,000	10,000
Activated carbon system		40,000	40,000
Ion exchange system		109,000	109,000
Electrowinning system		58,300	88,300
Atmospheric evaporator		4,000	0
Evaporator heaters		3,000	0
Evaporator scrubber		4,000	0
EQUIPMENT TOTAL		$ 339,000	$ 358,000
EQUIPMENT INSTALLATION		$ 168,000	$ 156,000
BUILDING MODIFICATIONS		131,000	131,000
PIPING		81,000	81,000
SUBTOTAL		$ 719,000	$ 726,000
ALLOWANCES FOR NONQUANTIFIED ITEMS			
Mobilization, bonding, insurance	5%	$ 36,000	$ 36,000
Painting	2%	14,000	15,000
Electrical, I&C	19%	137,000	138,000
SUBTOTAL		$ 906,000	$ 915,000

(cont.)

Table 8.3.   Continued

Category Items		LSA Elect. Alternative	HMT Elect. Alternative
CONTINGENCY	25%	$ 227,000	$ 229,000
SUBTOTAL		$1,133,000	$1,144,000
AREA ADJUSTMENT	15%	170,000	172,000
TOTAL ORDER-OF-MAGNITUDE ESTIMATE		$1,303,000	$1,316,000

Table 8.4.   Circuit Board Treatment Facility Operating Cost Estimate

Items	LSA Elect. Alternative	HMT Elect. Alternative
Labor (2,080 hr/yr @ $50/hr)	$104,000	$104,000
Chemicals	8,000	6,500
Electricity	8,000	9,000
Filters (52 sets/yr @ $38)	2,000	2,000
Activated carbon	1,600	1,600
Ion exchange resin	4,000	4,000
Maintenance	20,000	20,000
TOTAL ORDER-OF-MAGNITUDE ESTIMATE	$147,600	$147,100

Case Study 8.2: Recovery of Vanadium from an Electric Power Plant Wastewater

Background and Objectives.   The Roseton Station, located near Newburgh, New York, is a 1,200 megawatt, oil-fired electric power plant, operated by Central Hudson Gas & Electric Corporation (CHG&E). At the time of the project, the two units of the station burned a 1.5% sulfur content residual fuel oil with a vanadium content varying from 50 to 400 parts per million (ppm) of vanadium. A more detailed description of this project is presented in a paper by Schick and Rosain.[1]

Metal cleaning wastewaters are produced periodically from the following sources (in order of contribution to the waste load from highest to lowest):

1. air preheater wash

2. dust collector wash

3. economizer wash

4. I.D. fan and fan silencer wash

5. boiler fireside wash

6. stack wash

Annually, the air preheater wash accounts for 60% to 70% of the waste load, with the remainder being generated during major yearly shutdowns.

Waste characteristics vary widely, so an average composition is of little value. Ranges of values, however, are listed in Table 8.5. The waste would be characterized as having a high dissolved solids content, with high concentrations of magnesium, sulfate, vanadium, and nickel. Air preheater washes have the lowest pH at 2.5 to 3.5, and boiler fireside washes have a pH of about 8.5.

Table 8.5.   Power Plant Metal Cleaning Wastewater Characteristics

Parameter	Concentrations (mg/L)		
	Minimum	Maximum	Average
Dissolved solids (TDS)	5,000	250,000	15,000
Suspended solids (TSS)	2,000	20,000	
Magnesium	400	10,000	
Iron	5	3,000	
Vanadium	50	7,500	200
Nickel	1	850	
Sulfate	500	40,000	6,000
pH	2.5	8.5	

The plant's original wastewater treatment system consisted of a series of settling ponds that discharged to the Hudson River. Pond sludge was periodically dredged and sold to a reclaimer for vanadium recovery. Revised discharge permit conditions for vanadium and other heavy metals, as well as a desire to improve waste sludge handling, prompted CHG&E to initiate a major treatment system upgrade.

Process Description.   The design flow of wastewater was set at 1,200 gal/min, with the maximum volume of washwater produced during a washing event set at 4 million gal. Routine air preheater washes, occurring about every 2 to 4 weeks, generate 1 million gal each.

The chemistry of vanadium removal from wastewater is relatively complex. Vanadium is present in the raw plant wastewaters as a mixture of anionic and cationic oxide radicals, plus colloidal particles of the solid species V_2O_5 and V_2O_4. Extensive bench testing confirmed that the application of near stoichiometric quantities of ferrous sulfate at a pH of less than 6, followed by lime addition to a pH of 9.5, resulted in the optimal removal of vanadium and the other regulated heavy metals. Reaction is thought to be a combination of precipitation of ferrous vanadates, calcium vanadates, and physical adsorption on ferric hydroxide flocs. Lime, rather than caustic soda, was selected for pH adjustment because it produces a sludge that is more easily dewatered.

Figure 8.2 shows the process flow diagram for the treatment system. The waste is first degritted in a hydrocyclone and transported to one of two equalization basins with a combined capacity of 4 million gal. Waste is pumped to the reaction tank at a flow rate of about 200 gal/min. Liquid ferrous sulfate solution is injected into

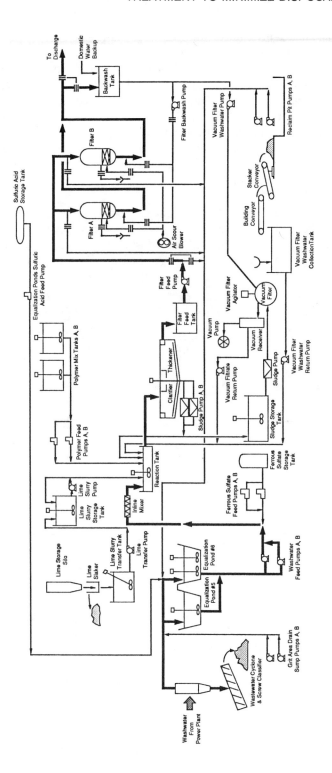

Figure 8.2. Roseton station's wastewater and sludge treatment facility.

the line and mixed prior to reaching the 4,000-gal reaction tank (20-min hydraulic retention). A 10% lime slurry is added to the reaction tank to control pH at 9.5.

The waste then flows by gravity to a 45-ft-diameter clarifier/thickener. A clarifier/thickener with an overflow rate of 0.1 gal/min/ft^2 was chosen because of the high solids loading on the clarifier. A portion of the sludge is returned to the reaction tank, maintaining solids at 2% to 3%, to seed the precipitation of calcium sulfate and to increase the overall floc density.

Clarifier effluent is filtered through dual media pressure filters and then discharged. Waste sludge from the clarifier is pumped to a mixed sludge storage tank. The sludge is then dewatered with a rotary drum filter.

Performance.   Table 8.6 shows typical plant performance factors for the operating facility. Dewatered sludge has averaged 26% solids with an average vanadium content of 17% as V_2O_5. The vanadium-rich sludge is loaded onto rail cars at a siding adjacent to the plant and sold to a reclaimer, turning what had been a waste disposal problem into an asset and, at the same time, yielding a cleaner plant effluent.[1]

Table 8.6.   Metal Cleaning Wastewater Treatment Performance

Parameter	Concentrations (mg/L)		
	Influent	Effluent	% Removal
pH	2.5	9.50	—
Iron	975	0.26	99.97
Copper	0.72	0.26	63.9
Chromium	0.31	0.04	87.1
Manganese	20	0.10	99.5
Zinc	19.6	0.04	99.8
Nickel	815	0.42	99.95
Vanadium	281	4.10	98.5
Lead	0.58	0.04	93.1

Case Study 8.3: Zero-Discharge Integrated Manufacturing Facility

Background and Objectives.   A company located a new manufacturing facility in a rural area. This facility was to be highly integrated, with a large number of the components of the product to be manufactured onsite.

Corporate management set minimizing the impact on the local environment as one of its objectives. In addition to environmental constraints, there was limited local water supply and wastewater disposal capacity.

Because of these factors, the company decided to investigate the feasibility of in-plant recycling of wastewaters, with the ultimate goal being zero discharge of wastewater from the site. Zero discharge has been achieved at large electric power generating stations for some time, but is not commonly attained at large integrated manufacturing facilities.

Water Management Computer Model

To assist in evaluating water and wastewater management alternatives, a computer model of the manufacturing facility's water use and wastewater production was developed. This water management model included water usage and quality requirements for more than 50 manufacturing processes in nine separate manufacturing areas, taking into account the chemical constituents and pollutants added to the water. In addition to the manufacturing processes, 20 water treatment processes were included in the model.

Treatment processes evaluated and modeled include:

- dissolved air flotation
- ultrafiltration
- pH adjustment with sulfuric acid
- pH adjustment with carbon dioxide
- pH adjustment with lime
- lime precipitation
- disinfection
- sludge dewatering
- spray drying
- free oil removal (API oil separator)
- ion exchange (anionic)
- ion exchange (calcium)
- ion exchange (sulfate)
- activated carbon adsorption
- filtration
- reverse osmosis
- vapor compression evaporation
- crystallization
- domestic water treatment
- biological treatment

Approaches to Zero Discharge

In order for zero discharge to be cost-effective, water use and wastewater production needed to be minimized. Methods used to achieve this goal include:

1. selecting manufacturing processes that use less water

2. increasing the recycle rate around traditional processes such as cooling towers

3. cascading the wastewater from one process to another for feed water, with little or no treatment

All three concepts were employed in developing a zero-discharge system for the facility.

To allow for cascade of used water, five water supplies, meeting varied water quality requirements, are provided at the facility. These systems are designated by W1 through W5 nomenclature. Table 8.7 defines these water supplies in terms of total dissolved solids (TDS), chemical oxygen demand (COD), and total suspended solids (TSS).

Table 8.7.   Integrated Manufacturer's Water Quality Requirements (in mg/L)

Water Quality	TDS	COD	TSS
W1—High purity process	0–10	0–1	0
W2—High quality process	10–150	0–10	0
W3—Softened process	100–150	0–10	0–10
W4—Unsoftened plant makeup	150–250	0–10	0–10
W5—Recycle process water	100–800	10–150	10–100

W1—Ultra pure process water
Ultra pure process water has an average TDS of 2 mg/L with a maximum of 10 mg/L. It is similar in quality to the effluent of an average two-bed ion exchange deminer-alizer. The major source of W1 water is evaporator distillate.

W2—High quality process recycle
W2 water consists of standard process water (W5 or local water) that has undergone TDS reduction using reverse osmosis. The high quality process recycle water has 10 to 150 mg/L TDS, zero TSS, and very low hardness. The water quality of this water is determined by the way the recycle system is operated.

W3—Softened local water supply
W3 water consists of potable water that is softened onsite to a hardness of about 50 mg/L.

W4—Potable water
Potable water consists of chlorinated filtered water as delivered in the city water dis-tribution system.

W5—Standard process recycle
Standard process recycle water consists of high quality recycle (W2) blended with secondary effluent from biological treatment that has been coagulated, settled, and filtered.

Approaches to achieving zero discharge include at-source treatment and recycle and end-of-plant treatment and recycle, or some combination of these in an integrated water management system.

At-Source Treatment and Recycle

This approach has appeal because waste streams are treated in small, possibly modular, treatment systems, and for only those parameters violating water reuse standards. In contrast, waste streams must be treated for all parameters in a large, centralized end-of-plant treatment system.

The at-source treatment alternatives examined focus on finding an appropriate treatment technology and reuse scheme for each individual waste stream or for logical groups of waste streams within each manufacturing unit. At-source treatment and recycle alternatives for two of the manufacturing units (Foundry and Paint Shop) are shown as examples.

The foundry recycle system is shown as Figure 8.3. Die casting is the primary contributor of emulsified oils. Ultrafiltration is recommended for this production unit. The majority of the other waste streams require treatment for reduction of TSS, BOD (biochemical oxygen demand), COD, and TDS. The ceramic mix area and laboratories only require treatment for TSS and TDS; however, their combined flow rate is approximately 1% of the total area flow. Therefore, they are combined with the other streams for treatment. A portion of the flow is used as W5 water after biological treatment and filtration. The remainder is routed through evaporation and activated carbon treatment to produce W1 water.

Two separate treatment systems are recommended for the wastewater generated at the Paint Shop (Figure 8.4). The building humidifier and spray booth humidifier waste streams require treatment for TSS and TDS only; therefore, filtration and reverse osmosis are provided for these waste streams. Parts wash and rinse, tool fixture wash, and paint booth washwater require treatment for TSS, BOD, COD, and TDS. Also, the paint booth washwater contains heavy metals. Under normal conditions, oil is not present in these waste streams. However, process upsets or dumps may result in oil discharge. Therefore, the following processes are provided: oil skimming and TSS removal in the equalization tank; heavy metals removal by hydroxide precipitation; biological oxidation for BOD and COD reduction; filtration for additional TSS reduction; evaporation for TDS reduction; and activated carbon for final organics reduction. Part of the treated water is diverted to the W5 storage tank after filtration, and the remaining water is directed to the evaporator to produce W1 water.

Conceptual level cost estimates for the at-source treatment systems are presented as Table 8.8. The capital cost was estimated at approximately $112 million, with an estimated annual operating cost of $9 million.

Integrated Water Management System

The water quality model was used to analyze numerous configurations of an end-

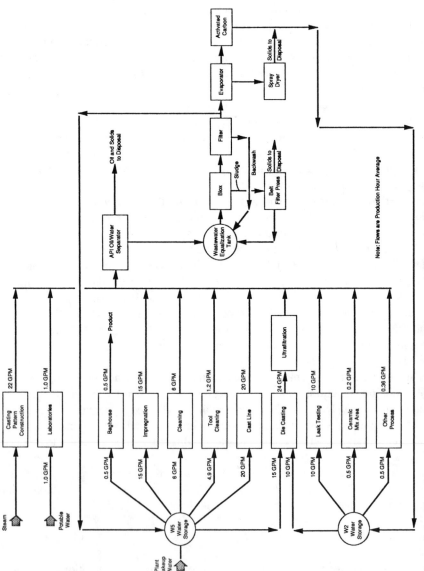

Figure 8.3.   Foundry water management plan for at-source treatment schematic.

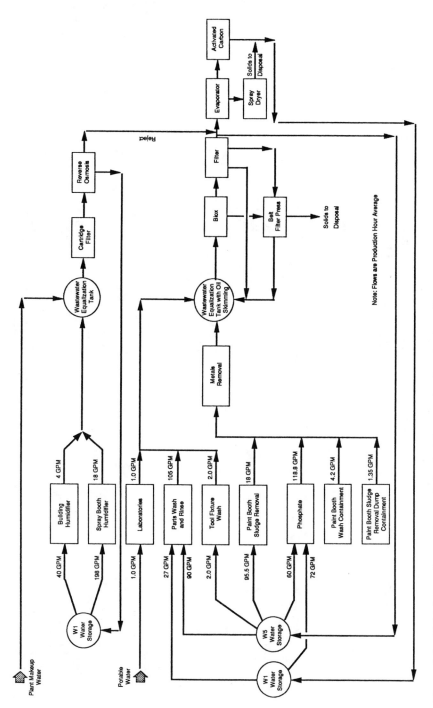

Figure 8.4.   Paintshop water management plan for at-source treatment schematic.

Table 8.8. Cost Estimate for At-Source Treatment System

Equipment Item	Size (gal)	Design Flow (gal/min)	No. Used	Estimated Installed Cost ($)
Foundry				
Ultrafiltration		17		530,000
Oil/water separator		150		207,000
Equalization	375,000			335,000
Biox		150		1,167,000
Emergency storage	1,875,000			881,000
Belt filter press				144,000
Filtration		150		458,000
Evaporator		45	2	3,206,000
Spray dryer		2		325,000
Activated carbon		90		541,000
W2 Storage	60,000			112,000
W5 Storage	60,000			112,000
Painting				
Wash & paint operations				
Equalization	750,000			700,000
Metals removal		250		589,000
Biox	5,000,000	375		2,022,000
Emergency storage				1,119,000
Belt filter press				144,000
Filter		375		693,000
Evaporator		138	2	6,280,000
Spray dryer		5		647,000
Activated carbon		275		1,057,000

(cont.)

Table 8.8. Continued

Equipment Item	Size (gal)	Design Flow (gal/min)	No. Used	Estimated Installed Cost ($)
Wash & paint operations cont'd				
W1 Storage	430,000			364,000
W5 Storage	162,000			203,000
Humidifiers				
Equalization	48,000			76,000
Cartridge filter		33		12,000
Reverse osmosis		33		111,000
W2 Storage	190,000			175,000
Manufacturing Area C				
Oil/water separator		8		36,000
Ultrafiltration		8		250,000
Equalization	124,000			173,000
Biox		44		559,000
Emergency storage	620,000			453,000
Belt filter press				144,000
Evaporator		22	2	2,087,000
Activated carbon		44		352,000
Spray dryer		2		325,000
W2 Storage	225,000			247,000
Manufacturing Area D				
Batch treat/equalization	1,300			8,000
DAF				72,000
Evaporator		2		336,000

(cont.)

Table 8.8. Continued

Equipment Item	Size (gal)	Design Flow (gal/min)	No. Used	Estimated Installed Cost ($)
Manufacturing Area D cont'd				
Activated carbon		2		75,000
Spray dryer		2		325,000
W2 Storage	5,000			11,000
Reverse osmosis		5		35,000
Manufacturing Area E				
Batch treat/equalization	1,000			8,000
DAF				72,000
Evaporator		2		336,000
Activated carbon		2		75,000
Spray dryer		2		325,000
W1 Storage	1,000			8,000
Manufacturing Area F				
Batch treat/equalization	2,600			13,000
DAF				72,000
Evaporator		4		509,000
Spray dryer		2		325,000
Activated carbon		4		84,000
W2 Storage	6,800			23,000
Reverse osmosis		6		38,000
Manufacturing Area G				
Equalization	15,000			39,000
Evaporator		15		1,126,000

(cont.)

Table 8.8.   Continued

Equipment Item	Size (gal)	Design Flow (gal/min)	No. Used	Estimated Installed Cost ($)
Manufacturing Area G cont'd				
Spray dryer		2		325,000
Activated carbon		15		185,000
W5 Storage	10,000			30,000
Equalization	15,000			39,000
Cartridge filter		15		6,000
Activated carbon		15		185,000
W5 Storage	10,000			30,000
W2 Storage	500			7,000
Powerhouse				
Cooling tower				
Lime softening		900		430,000
Cooling tower blowdown and power plant				
Oil/water separator		260		68,000
Equalization	250,000			263,000
Evaporator		130	2	6,059,000
Spray dryer		8		850,000
W1 Storage	238,000			255,000
W5 Storage	1,305,000			709,000
Miscellaneous Systems				
Equalization	3,500			15,000
Evaporator		4		509,000
Spray dryer		2		325,000

(cont.)

Table 8.8.   Continued

Equipment Item	Size (gal)	Design Flow (gal/min)	No. Used	Estimated Installed Cost ($)
Miscellaneous Systems cont'd				
Activated carbon	3,500			72,000
W5 Storage				15,000
Domestic Wastewater Treatment				677,000
TOTAL ESTIMATED EQUIPMENT COST				41,805,000
Piping (20% of modified equipment)				10,451,250
Electrical (7% of modified equipment)				3,146,613
Instrumentation and control (10% of modified equipment)				4,645,000
Site work (10% of modified equipment)				4,645,000
Mobilization and insurance (7% of modified equipment)				3,146,613
TOTAL ESTIMATED CONSTRUCTION COSTS				67,839,476
Scope Completion Allowance (30% of construction costs)				20,351,843
Subtotal				88,191,319
Engineering (15% of subtotal)				13,228,698
Subtotal				101,420,016
Contingency (10% of subtotal)				10,142,002
TOTAL CONCEPTUAL LEVEL CAPITAL COSTS				111,562,018

of-plant wastewater treatment facility. It was quickly determined that some at-source treatment for oil removal and metals removal was cost-effective. The selected integrated treatment system is presented in Figure 8.5. This treatment system contains the following treatment process units: oil and grit separation, biological oxidation, filtration, vapor compression evaporation, crystallization, and granular activated carbon adsorption. Lime softening is provided for power plant cooling tower makeup to allow for high cycles of concentration. In addition to the centralized treatment system, at-source treatment processes are provided, including ultrafiltration for foundry wastewaters and metals removal for paint shop wastewaters.

The treatment train receives all process effluents from the business units, treats the wastewaters, and recycles the water back to the business units, along with makeup water from a local water treatment plant. Domestic wastewater is treated separately and disposed of by land application.

Vapor compression evaporation was recommended as the primary TDS reduction process because this equipment is less susceptible to process upsets than reverse osmosis (RO). Additional pretreatment would also be required to protect RO from organic fouling.

Conceptual-level cost estimates for this integrated water management system are presented as Tables 8.9 and 8.10. The capital cost was estimated at approximately $62 million, with an estimated annual operating cost of $6 million. Further analysis showed that if dry cooling towers were adopted by the power plant, lime softening of the cooling tower makeup would reduce capital costs to $54 million and annual O&M costs to $5 million. These costs are between half and two-thirds of the costs of at-source treatment and recycle.

SEGREGATED TREATMENT TO REDUCE HAZARDOUS MIXTURES

In the past, it has been cost-effective to operate a centralized industrial waste treatment plant rather than to provide individual treatment systems for each production unit in a large manufacturing facility. However, hazardous waste regulations have classified the residues from the treatment of wastes from certain industrial operations (i.e., electroplating or chromate conversion coating of aluminum) as listed hazardous wastes, regardless of their actual composition or toxicity. This is complicated by the RCRA mixture rule that states that mixing any quantity of a hazardous waste with a nonhazardous waste renders the entire mixture hazardous.

The combination of these two regulations makes it imperative that a manufacturer investigate whether combined treatment of all waste streams makes sense. It is often economical to provide treatment for hazardous waste streams separately from all other wastes, especially when the hazardous streams make up only a small portion of total waste flow in the facility or contribute to producing only a small volume of sludge when treated.

Following is a case study in which an aerospace manufacturing facility plans to provide separate treatment (and recycle) of the wastewater from chromate conversion coatings operations. This segregation of wastes will result in the industrial

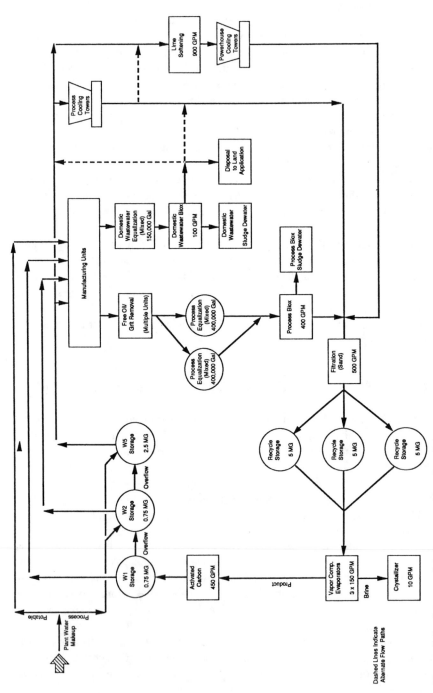

Figure 8.5. Integrated water management system.

Table 8.9. Capital Cost Estimate for Integrated Water Management System

Equipment Used	Size (gal)	Design Flow (gal/min)	No. Used	Estimated Installed Cost ($)
Chemical feed systems			3	144,000
Ultrafiltration			2	780,000
Oil/water separators				333,000
Metals removal—lime prec.		250		589,000
Process equalization (mixed)	400,000		2	697,000
Process biological treatment		400		2,102,000
Domestic biological treatment	150,000	100		677,000
Lime-soda softening		900		430,000
Recycle storage	5,000,000		3	3,358,000
Filtration		500		824,000
Vapor compression evaporation				
with crystallizer		150	3	9,904,000
Activated carbon		450		1,421,000
Plant water storage			3	1,193,000
Building				379,000
TOTAL ESTIMATED EQUIPMENT COSTS				22,831,000
Piping (20% of modified equipment)				5,707,750
Electrical (10% of modified equipment)				2,536,778
Instrumentation and control (7% of modified equipment)				1,718,462
Site work (10% of modified equipment)				2,536,778
Mobilization and insurance (7% of all above)				2,473,154
Subtotal				37,803,922

(cont.)

Table 8.9.   Continued

Equipment Used	Size (gal)	Design Flow (gal/min)	No. Used	Estimated Installed Cost ($)
Scope allowance (30% of subtotal)				11,341,177
TOTAL ESTIMATED CONSTRUCTION COSTS				49,145,098
Engineering (15% of construction costs)				7,371,765
Subtotal				56,516,863
Contingency (10% of construction costs)				5,651,686
TOTAL CONCEPTUAL LEVEL CAPITAL COSTS				62,168,549

Table 8.10. O&M Cost Estimate for Integrated Water Management System

Operation and Maintenance Items	Units	Unit Cost ($)	Annual Usage	Annual Cost ($)
Maintenance (4% of equipment)	lump sum			913,240
Labor—adm. and tech.	man-years	65,000	9	585,000
Labor	man-years	45,000	20	900,000
Electricity demand charge	kW	156.96	3,653	538,000
Electricity energy charge	1,000 kWh	20.30	18,842	375,000
Lime	tons	100	123	12,300
Sulfuric acid	tons	100	135	13,500
Carbon	lb	1	376,145	376,145
Polymer	gal	5	1,000	5,000
Degreaser	gal	2	42,000	84,000
Lime sludge disposal	tons	125	633	79,125
Filtration sludge disposal	tons	125	772	96,500
Biological sludge disposal	tons	125	294	36,750
Oil sludge disposal	tons	125	2,821	352,625
Crystallizer sludge disposal	tons	125	661	82,625
Precoat	tons	600	129	77,400
Ferric chloride	tons	500	8	4,000
Water	1,000 gal	3.50	351,101	1,228,853
TOTAL ANNUAL CONCEPTUAL LEVEL O&M COSTS				5,760,063

wastewater treatment plant sludges becoming classified as nonhazardous. Cutting down on the amount of wastes to be disposed of justifies the construction and operation of two separate treatment plants.

Case Study 8.4: Separate Treatment of a Chromate Conversion Coatings Waste

Background and Objectives.   CH2M HILL performed a waste minimization study for a major aerospace manufacturing facility in the western United States. This study identified rinsewaters from the chromate conversion coating (Iriditing) of aluminum as the waste stream on which to concentrate reduction efforts, since this waste is responsible for the sludge from the entire industrial wastewater treatment plant being classified as hazardous. These rinsewaters (approximately 15,000 gal/day) are mixed with other industrial wastes (212,000 gal/day) for treatment at a central industrial wastewater treatment plant.

The EPA lists "wastewater treatment sludges from the chemical conversion coating of aluminum" as hazardous under the classification F019, specifically because they typically contain hexavalent chromium or cyanide (RCRA Appendix VII). In addition to being used in the chromium conversion coating process, hexavalent chromium compounds are used at the facility to deoxidize surfaces (remove oxide surface coating) when preparing parts for chromate conversion coating. Table 8.11 shows the amounts of chromium discharged to the waste treatment plant from the three locations that employ chromate conversion coating at the facility. These data were estimated from composition of the process baths involved and from approximations by company personnel of the production loads on these processes and typical drag-out rates.

Table 8.11.   Chromium Discharges from Conversion Coating Facilities

Location	Discharge (lb/day)
Chem Mill Facility	0.90
Remote Location A	0.05
Remote Location B	0.01
Total	0.96

In 1987, 170 tons of dewatered sludge were produced at the industrial wastewater treatment facility. The waste did not contain sufficient toxic heavy metals to be classified as a "characteristic" hazardous waste, but since it was a listed waste, it was disposed of as hazardous at a cost of $210,000. Thus, less than 1 lb of chromium discharged to the treatment plant resulted in more than 1,000 lb of sludge per day being classified as hazardous. Hence, there was significant incentive to eliminate these chromium-containing wastewaters from the industrial treatment plant.

Eliminating the F019 classification of the industrial wastewater treatment sludge could be achieved only by attaining zero discharge of rinsewaters and baths from the chromate conversion coating and the deoxidation processes. To achieve this objective, a closed-loop rinsewater system would have to be installed, with all residues being hauled offsite for disposal as hazardous wastes.

Discussion of Alternatives.   Four unit processes were considered necessary to implement a closed-loop system:

1. rinsewater recycle

2. chemical recovery

3. volume reduction

4. bath purification

Rinsewater recycle is a necessity since rinsewater is usually maintained at a low concentration of contamination to effect good rinsing. The existing operation produced approximately 15,000 gal of waste rinsewater per day (125,000 lb) in the process of removing the 1 lb of chromium per day. Unit processes considered for rinsewater cleanup included:

- ion exchange

- reverse osmosis

- ion transfer membranes

- electrodialytic processes

Recovering chromium from the rinsewater would eliminate the major toxic constituent from this waste stream and simplify disposal of any blowdown stream. Processes considered for process chemical recovery included:

- ion exchange

- electrodialytic processes

Volume reduction was considered necessary to reduce the size of the rinsewater recycle system and also to reduce the quantity of blowdown required to prevent build-up of salts in the system. Methods for volume reduction included:

- innovative rinsing

- evaporation

Bath purification is necessary if process chemicals are to be recovered from the rinsewater and returned to the bath, since such systems will return contaminants as well as useful chemicals. Otherwise, impurities are concentrated in the bath until the bath is no longer functional, necessitating disposal of a large quantity of hazardous material. It makes little sense to recover 1 lb of chromium per day if this recovery necessitates the disposal of a 20,000-gal tank of Iridite solution containing over 800 lb of chromium. Also, since impurities are present in the process solutions in much higher concentrations than in the rinsewater, bath purification is simpler than rinsewater cleanup. Because cations are removed in the bath and are not carried over to the rinsewater, the volume of cation regenerant is significantly reduced. Technologies evaluated for purification of chrome-containing process solutions included:

- electrodialytic processes

- porous pot

- ion exchange

These technologies were evaluated for potential for waste reduction, effects on production, and relative cost. Existing users of the technologies were contacted to discuss their operating experience. Site visits were made to collect information on the most promising technologies.

Estimating Waste Loads.   Compositions of process baths containing chromium were measured. Normal production rates were received from production personnel and used to predict volumes of process solutions dragged out each day (using 1 gal of drag-out per 1,000 ft² of parts, which were predominantly flat sheets of aluminum). This information, along with composition of the process solutions, was used to estimate drag-out of individual ions to the rinsewater. The results of this analysis are presented as Table 8.12.

The small amount of the chromium produced in the conversion coating (Iridite) process that is discharged as waste in this facility is somewhat surprising. Approximately 31 g (.07 lb) per day of chromium is dragged out of the Iridite, compared with a total of 427 g (.94 lb) per day for all sources. The principal source of chrome is drag-out from the deoxidation tanks. The drag-out rate is higher because the chromate concentration is higher in the deox tanks, and the parts undergo numerous immersions in the deox as they are processed, whereas they are only processed once in the Iridite.

In addition, the deox solutions contain high concentrations of nitric acid, which is used to dissolve the aluminum oxide layer that forms on the surface of the metal when the aluminum surface is exposed to air. Thus the deox drag-out contains high concentrations of nitrate, which need to be accommodated in the closed-loop rinsing chromium recovery system.

Recommended System.   Ion exchange (IX) was recommended for rinsewater cleanup. IX is an established technology and can also segregate chromic and nitric

Table 8.12.   Estimated Drag-Out Rates from Conversion Coating Facilities

Ion	CM-3	CM-12	CM-15	CM-22	CM-27	CM-30	A-2	A-4	B	Total	Total Less Deox
Al	126.9	8.1	.15	—	.62	.01	1.03	.02	.32	137.15	.5
Cd	—	—	.01	—	—	0	—	0	0	.01	.01
Ca	—	—	.12	—	—	.01	—	.02	.26	.41	.41
Cr(3)	108.1	16.2	5.1	.81	1.24	.39	2.05	.03	11.01	144.93	17.34
Cr(6)	244.4	16.5	4.5	2.43	1.27	.35	2.09	1.03	9.69	282.26	18.6
Cr(T)	352.5	32.7	9.6	3.24	2.51	.74	4.14	1.06	20.7	427.19	35.94
Cu	—	.6	0	—	.03	0	.05	0	0	.68	0
Pb	—	—	0	—	—	0	—	0	0	0	0
Hg	—	—	0	—	—	0	—	0	0	0	0
Ni	—	—	0	—	—	0	—	0	0	0	0
Ag	—	—	0	—	—	0	—	0	0	0	0
Zn	1.9	.12	0	—	.01	0	.02	—	0	2.05	0
NO_3	8700	555	—	—	—	—	351	—	—	9606	0
Na	155.1	14.4	4.2	1.4	1.1	.3	1.8	.5	9.1	187.9	15.8

All values in grams/day

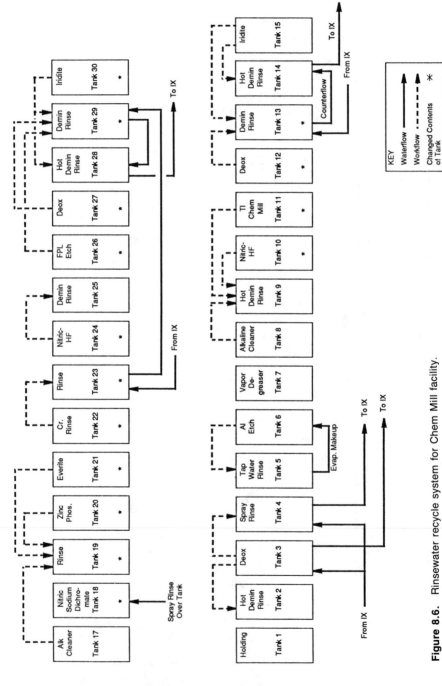

Figure 8.6. Rinsewater recycle system for Chem Mill facility.

acid from cations and concentrate them for reuse. Reduced rinse flows are accomplished by providing some counterflow rinsing, so that four individual recycle streams are used for the seven tanks. A schematic of the rinsewater recycle system is shown as Figure 8.6. An electrodialytic bath maintenance unit was recommended for the principal deox bath (Figure 8.7). Finally, evaporation was recommended for concentration of ion exchange regenerant solutions, either to enhance recovery or to reduce cost of disposal.

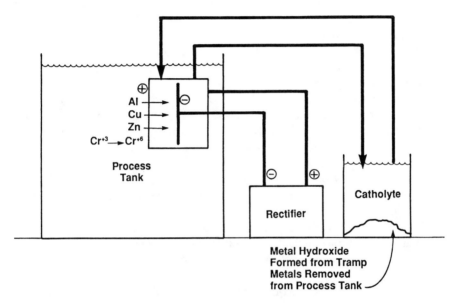

Figure 8.7. Iridite and deoxidizer bath maintenance system.

A process flow diagram for the rinsewater treatment system is shown in Figure 8.8. Contaminated rinsewater is passed through a cartridge filter to remove parti-

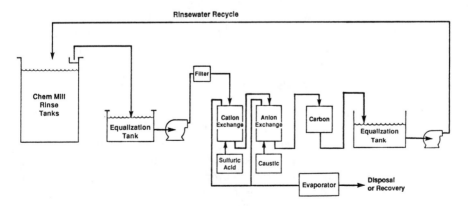

Figure 8.8. Rinsewater recycle system for Chem Mill facility.

cles that could plug the ion exchange resin. A cartridge filter was recommended since suspended solids loading is low in these acidic rinsewaters. Also, a cartridge filter does not require backwashing, so less wastewater is generated. Cation exchange is used to remove metallic impurities such as trivalent chromium, aluminum, sodium, and zinc. Then an anionic exchange resin is used to remove hexavalent chromium and nitrate ions. Finally, the rinsewater is treated with activated carbon to remove traces of organics that could foul ion exchange resins if allowed to build up in the closed-loop rinsewater system. The carbon is placed after, rather than in front of, the ion exchange resin, since carbon removes hexavalent chromium, reducing the potential for chromium recovery.

The cation exchange resin is regenerated with sulfuric acid, removing the metals and returning the resin to its acid form. The volume of this acidic regenerant solution is reduced by evaporation and disposed of offsite.

The anion exchange resin is regenerated with caustic soda (sodium hydroxide). The resulting regenerant solution is a caustic mixture of sodium chromate, sodium nitrate, and sodium hydroxide. This regenerant is then concentrated by evaporation. A discussion of potential recovery of nitric acid and chromic acid is included later.

Because of the low production of contaminants at Remote Sites A and B, it is not cost-effective to provide these locations with completely independent demineralizer units.

It is more cost-effective to provide these locations with portable ion exchange units, which are to be returned to the Chem Mill facility for regeneration on a scheduled basis. A typical portable ion exchange unit is shown as Figure 8.9.

Design Criteria and Major Equipment Sizing.   The anion and cation loadings for the central and remote rinsewater ion exchange were estimated from design workloads, bath composition, and predicted drag-outs. Worksheets for ion exchange sizing and projections of regenerant production are provided in Tables 8.13 through 8.15.

The major cation in these rinsewaters is the hydronium ion. Since the cation resin is to be used for demineralization and is regenerated to the acid form, only non-hydronium cations are removed from the rinsewater and used to calculate loading on the resin. The significant anions are nitrate (from the nitric acid used in the deox baths) and hexavalent chromium (assumed to be in the form of chromate ions for this analysis). The pH of the combined rinsewater is computed based on a rinse flow of 15 gal/min for the Chem Mill system and 4 gal/min for the remote systems at Sites A and B.

The sizes of the ion exchange beds were determined from the estimated daily ionic loading on the cation and anion resins. Sizing was based on a hydraulic loading of 2 gal/min/ft^3 of resin and a regeneration frequency of not more than once per day for the central system and once per week for the remote sites. Based on typical regenerant chemical usage (3 lb of sodium hydroxide per cubic foot for anion resin and 10 lb of sulfuric acid per cubic foot for the cation resin), the compositions of

with
MAGNETIC PUMP

with **IN-TANK PUMP**

Figure 8.9. Remote ion exchange unit for Chem Mill facility (Photo courtesy of Sethco Manufacturing Corporation, Freeport, New Jersey).

the regenerant solutions were estimated for evaluation of recovery potential or disposal.

The combined regenerant composition was determined based on the flow-weighted average of the individual regenerant solutions (Table 8.16). The effect of mixing the acidic cation resin regenerant (H + ions) with the caustic anion resin regenerant (OH − ions) was accounted for in the production of water and the net caustic pH. The combined regenerant solution would be a caustic mixture of sodium salts of sulfate, nitrate, and chromate. Evaporation to 25% salt content would reduce the disposal volume to 28 gal/day.

Table 8.13.   Worksheet for Chem Mill Ion Exchange System

Cations	MW	Charge	CM-3 Deox	CM-12 Deox	CM-15 Iridite	CM-22 Rinse	CM-27 Deox	CM-30 Iridite	Total Drag-Out/Day (Grams)	(Moles)	(Equiv)
Al(+3)	27	3	127	8.1	0.2	—	0.6	—	136	5.0	15.1
Cr(+3)	52	3	108	16.2	5.1	0.8	1.2	0.4	132	2.5	7.6
Zn(+2)	65.4	2	2	0.1	—	—	—	—	2	0.0	0.1
Na(+1)	23	1	155	14.4	4.2	1.4	1.1	0.3	176	7.7	7.7
H(+1)	1	1							129	129.2	129.2
TOTAL CATIONS									575	144.5	159.6
CATIONS WITHOUT H+									446	15.3	30.4
Anions											
CrO$_4$(−2)	52	2	244	16.5	4.5	2.4	1.3	0.4	269	5.2	10.4
NO$_3$(−1)	62	1	8,700	555.0	—	—	—	—	9,255	149.3	149.3

TOTAL ANIONS

Average rinsewater pH = 2.8

Cation resin loading (eq/day) =	30.4	
IR-120 capacity (Kgrains/ft³) =	27.3	
Eq/Kgrain =	1.3	
Exchange capacity (eq/ft³) =	35.4	
Resin use (ft³/day) =	0.9	
Bed volume (ft³) =	10.0	
Est. days between regen. =	11.6	
Design days between regen. =	7.0	

Cation regenerant =	H$_2$SO$_4$
H$_2$SO$_4$ load (lb/ft³) =	10
H$_2$SO$_4$ used per regen. (lb) =	100
H$_2$SO$_4$ (eq/lb) =	9
Equivalents of regen. =	927
H$_2$SO$_4$ volume (bed volumes) =	5
H$_2$SO$_4$ volume (gal) =	374
H$_2$SO$_4$ volume (L) =	1,414

(cont.)

Table 8.13. Continued

COMPOSITION OF CATIONIC RESIN REGENERANT

Cations	Equiv	MW	Charge	Amount (grams)	Conc. (g/L)	Production (g/day)
Al(+3)	106	27	3	951	0.67	136
Cr(+3)	53	52	3	922	0.65	132
Zn(+2)	0	65.4	2	14	0.01	2
Na(+1)	54	23	1	1,235	0.87	176
H(+1)	714	1	1	714	0.50	102
Anions						
SO$_4$ (1−2)	927	98	2	45,400	32.11	6,486
TOTAL	1,853			49,236	34.83	7,034

Anion resin loading (eq/day) =   154.45   Anion regenerant =   NaOH
IRA-93 capacity (Kgrains/ft³) =   21.00   NaOH load (lb/ft³) =   3.0
Eq/Kgrain =   1.30   NaOH used per regen. (lb) =   30.0
Exchange capacity (eq/ft³) =   27.24   NaOH (eq/lb) =   11.4
Resin use (ft³/day) =   5.67   Equivalents of regen. =   340.5
Bed volume (ft³) =   10.0   NaOH volume (bed volumes) =   5.0
Est. days between regen. =   1.76   NaOH volume (gal) =   374
Design days between regen. =   1   NaOH volume (L) =   1,414

(cont.)

Table 8.13.   Continued

COMPOSITION OF ANIONIC RESIN REGENERANT

Anions	Equiv	MW	Charge	Amount (grams)	Conc. (g/L)	Production (g/day)
$CrO_4(-2)$	10.4	52	2	269	0.19	269
$NO_3(-1)$	149.3	62	1	9,255	6.55	9,255
$DH^3(-1)$	180.9	17	1	3,075	2.18	3,075
Cations						
$Na(+1)$	340.5	23	1	7,832	5.54	7,832
TOTAL	681			20,430	14.45	20,430

Table 8.14.   Worksheet for Remote Site A Ion Exchange System

Cations	A-2 Deox	A-4 Iridite	(Grams)	Total Drag-Out/Day (Moles)	(Equiv)
$Al(+3)$	1.03	0.02	1.05	0.04	0.12
$Cr(+3)$	2.05	0.03	2.08	0.04	0.12
$Zn(+2)$	0.02	—	0.02	0.00	0.00
$Na(+1)$	1.80	0.50	2.30	0.10	0.10
$H(+1)$			5.78	5.78	5.78
TOTAL CATIONS			11.23	5.96	6.12
CATIONS WITHOUT H+			5.45	0.18	0.34

(cont.)

Table 8.14.   Continued

Anions

$CrO_4(-2)$	2.09	1.03	3.12	0.06	0.12
$NO_3(-1)$	351.00	—	351.00	5.66	5.66

TOTAL ANIONS

Average rinsewater pH = 3.28

Cation resin loading (eq/day) =	0.34	Cation regenerant =	H_2SO_4
IR-120 capacity (Kgrains/ft³) =	27.30	H_2SO_4 load (lb/ft³) =	10.0
Eq/Kgrain =	1.30	H_2SO_4 used per regen. (lb) =	18.0
Exchange capacity (eq/ft³) =	35.41	H_2SO_4 (eq/lb) =	9.3
Resin use (ft³/day) =	0.01	Equivalents of regen. =	166.8
Bed volume (ft³) =	1.80	H_2SO_4 volume (bed volumes) =	5.0
Est. days between regen. =	189	H_2SO_4 volume (gal) =	67
Design days between regen. =	90	H_2SO_4 volume (L) =	254

COMPOSITION OF CATIONIC RESIN REGENERANT

Cations	Equiv	MW	Charge	Amount (grams)	Conc. (g/L)	Production (g/day)
Al(+3)	11	27	3	95	0.37	1
Cr(+3)	11	52	3	187	0.74	2
Zn(+2)	0	65.4	2	2	0.01	0
Na(+1)	9	23	1	207	0.81	2
H(+1)	136	1	1	136	0.54	2

(cont.)

Table 8.14.   Continued

Anions

SO$_4$ (−2)	167	98	2	8,172	32.11	91
TOTAL	334			8,799	34.5	98

Anion resin loading (eq/day) = 5.78   Anion regenerant = NaOH
IRA-93 capacity (Kgrains/ft^3) = 21.00   NaOH load (lb/ft^3) = 3.0
Eq/Kgrain = 1.30   NaOH used per regen. (lb) = 5.4
Exchange capacity (eq/ft^3) = 27.24   NaOH (eq/lb) = 11.4
Resin use (ft^3/day) = 0.21   Equivalents of regen. = 61.3
Bed volume (ft^3) = 1.80   NaOH volume (bed volumes) = 5.0
Est. days between regen. = 8.5   NaOH volume (gal) = 67
Design days between regen. = 7   NaOH volume (L) = 254

COMPOSITION OF ANIONIC RESIN REGENERANT

Anions	Charge	MW	Equiv	Amount (grams)	Conc. (g/L)	Production (g/day)
CrO$_4$(−2)	2	52	0.8	22	0.09	3
NO$_3$(−1)	1	62	39.6	2,457	9.66	351
OH^3(−1)	1	17	20.8	354	1.39	51
Cations						
Na(+1)	1	23	61.3	1,410	5.54	201
TOTAL			123	4,242	16.67	606

Table 8.15.   Worksheet for Remote Site B Ion Exchange System

Cations	(Grams)	Total Drag-Out/Day (Moles)	(Equiv)
Al(+3)	0.32	0.01	0.04
Cr(+3)	11.01	0.21	0.64
Zn(+2)	0.00	0.00	0.00
Na(+1)	9.10	0.40	0.40
TOTAL CATIONS	20.43	0.62	1.07
Anions			
$CrO_4(-2)$	9.69	0.19	0.37
$NO_3(-1)$	0.00	0.00	0.00
$OH^3(-1)$	11.79	0.69	0.69
TOTAL ANIONS	21.48	0.88	1.07
ANIONS WITHOUT OH–			0.37

Average rinsewater pH = 9.80

Cation resin loading (eq/day) =	1.07	Cation regenerant =	H_2SO_4
IR-120 capacity (Kgrains/ft³) =	27.30	H_2SO_4 load (lb/ft³) =	10.0
Eq/Kgrain =	1.30	H_2SO_4 used per regen. (lb) =	18.0
Exchange capacity (eq/ft³) =	35.41	H_2SO_4 (eq/lb) =	9.3
Resin use (ft³/day) =	0.03	Equivalents of regen. =	166.8
Bed volume (ft³) =	1.80	H_2SO_4 volume (bed volumes) =	5.0
Est. days between regen. =	60	H_2SO_4 volume (gal) =	67
Design days between regen. =	30	H_2SO_4 volume (L) =	254

(cont.)

Table 8.15.   Continued

COMPOSITION OF CATIONIC RESIN REGENERANT

Cations	Equiv	MW	Charge	Amount (grams)	Conc. (g/L)	Production (g/day)
Al(+3)	1	27	3	10	0.04	0
Cr(+3)	19	52	3	330	1.30	11
Zn(+2)	0	65.4	2	0	0.00	0
Na(+1)	12	23	1	273	1.07	9
H(+1)	135	1	1	135	0.53	4
Anions						
SO$_4$ (−2)	167	98	2	8,172	32.11	272
TOTAL	334			8,920	35.05	297

Anion resin loading (eq/day) = 0.37
IRA-93 capacity (Kgrains/ft^3) = 21.00
Eq/Kgrain = 1.30
Exchange capacity (eq/ft^3) = 27.24
Resin use (ft^3/day) = 0.01
Bed volume (ft^3) = 1.80
Est. days between regen. = 131.6
Design days between regen. = 90

Anion regenerant = NaOH
NaOH load (lb/ft^3) = 3.0
NaOH used per regen. (lb) = 5.4
NaOH (eq/lb) = 11.4
Equivalents of regen. = 61.3
NaOH volume (bed volumes) = 5.0
NaOH volume (gal) = 67
NaOH volume (L) = 254

(cont.)

Table 8.15. Continued

COMPOSITION OF ANIONIC RESIN REGENERANT

	Equiv	MW	Charge	Amount (grams)	Conc. (g/L)	Production (g/day)
Anions						
$CrO_4(-2)$	33.5	52	2	872	3.43	10
$NO_3(-1)$	0.0	62	1	0	0.00	0
$OH^3(-1)$	27.7	17	1	472	1.85	5
Cations						
$Na(+1)$	61.3	23	1	1,410	5.54	16
TOTAL	123			2,753	10.82	31

Table 8.16.   Projected Combined Regenerant Composition

Ions	MW	Charge	Central System Cation	Central System Anion	Site A Cation	Site A Anion	Site B Cation	Site B Anion	Combined (g/day)	Combined (eq/day)	Net (g/day)	Net (eq/day)	Net (g/L)
						Contributions (g/day)							
Al(+3)	27	3	136		1		0		137	15.3	137	15.3	0.08
Cr(+3)	52	3	132		2		11		145	8.4	145	8.4	0.08
Zn(+2)	65.4	2	2		0		0		2	0.1	2	0.1	0.00
Na(+1)	23	1	176	7,832	2	201	9	16	8,236	358.1	8,236	358.1	4.64
H(+1)	1	1	102		2		4		108	107.9	0	0.0	0.00
SO₄(-2)	98	2	6,486		91		272		6,849	139.8	6,849	139.8	3.86
CrO₄(-2)	52	2		269		3		10	282	10.8	282	10.8	0.16
NO₃(-1)	62	1		9,255		351		0	9,606	154.9	9,606	154.9	5.41
OH(-1)	17	1		3,075		51		5	3,131	184.2	1,296	76.2	0.73
TOTAL			7,034	20,430	98	606	297	31	28,496	979	26,553	764	14.95
VOLUME (gal) =			53	374	1	36	2	1	468				
VOLUME (L) =			203	1,421	3	138	9	3	1,777				
ph = 12.6													

Segregated Treatment Cost Estimates.   Projected capital and operating costs for the ion exchange systems (a central system with regeneration facilities and two remote systems) are provided in Tables 8.17 and 8.18. The cost of installing this system is estimated at approximately $294,000 and would result in a savings of $148,300 per year, with a payback period of less than 2 years for this investment.

Installation of a bath maintenance system on the Chem Mill Tank 3 deox tank would cost approximately $16,000. This would decrease cation loads to the demineralizer system by approximately 90% from this source, reducing overall cation loading to the demineralizer system by approximately 80%. The annual savings from installing a bath maintenance system for this tank are provided in Table 8.19. This table shows an annual savings of $4,400, for a projected payback period of approximately 3.6 years.

An additional production benefit not included in this analysis is improved consistency and increased operating life of the process solution, which reduces the need for disposal. The nonquantified production benefits (and low capital cost) were sufficient to convince management to adopt this improvement, despite its relatively long projected payback period.

Table 8.17.   Ion Exchange System Capital Cost Estimate

Item	Cost (1988 $)
Central IX system	40,000
Cartridge filters with feed pumps	6,600
Discharge pumps (2)	3,000
Equalization tanks (2-1,000 gal FRP)	9,200
Spray rinse storage tank (1,000 gal FRP)	4,500
Carbon adsorption (15 ft^3)	4,500
Remote IX canisters (4 sets)	2,000
Remote pumps (2)	2,000
Hydrostat surge tank (1,500 gal FRP)	6,000
Remote cartridge filters (2)	1,000
Remote carbon adsorption tanks (2)	3,000
Anion exchange resin (18 ft^3 @ $100/ft^3)	1,400
Cation exchange resin (18 ft^3 @ $80/ft^3)	1,400
Evaporators (2-15 gal/hr capacity)	9,500
Immersion heaters (2)	1,000
Subtotal equipment cost	105,000
Installation labor (640 hr @ $39.70/hr)	25,400
Allowance for nonquantified items (mechanical, 10%; I&C, 5%; electrical, 8%; and miscellaneous, 10% = 33% of constructed cost before contingency)	64,500
Estimated constructed cost	195,500
Contingency (25% of estimated constructed cost)	48,500
Engineering	50,000
ESTIMATED TOTAL INSTALLATION COST	$294,000

Table 8.18.   Ion Exchange System Estimated Operating Costs and Payback

Item	Cost (1988 $)
1. Anion regenerant chemicals (8,000 lb NaOH)	2,400
2. Cation regenerant chemicals (5,500 lb H_2SO_4)	400
3. Power for evaporation ($.10/gal)	10,000
4. Other power (10 hp)	3,800
5. Disposal cost (10,000 gal @ $1/gal)	10,000
6. Carbon use (1 change/year)	1,100
7. Filter cartridge replacement and disposal	2,000
8. Labor (1,000 hr @ $11.00/hr)	11,000
9. Annual operating cost	40,700
10. Reduction in sludge disposal cost	189,000
11. Annual savings	148,300
12. System installed cost	294,000
13. Payback period (years)	2

Table 8.19.   Estimated Cost Savings Due to Bath Maintenance System

Item	Cost (1988 $)
1. Regenerant chemicals (4,400 lb H_2SO_4)	300
2. Power for bath maintenance (1 kW)	(600)
3. Evaporation power (12,000 gal)	1,200
4. Disposal cost (20% reduction)	2,000
5. Bath chemicals savings	1,500
6. Total cost savings per year	4,400
7. Installed cost of system	16,000
8. Payback period (years)	3.6

Another modification that was considered was an electrodialytic chromic acid recovery system for the anion regeneration solutions (Figure 8.10). Use of this system would significantly reduce the waste disposal volume and risks associated with disposal of chromium-containing wastes. Annual savings were projected at approximately $10,000 (Table 8.20) for a system with an estimated installed cost of $40,000, with a payback period of approximately 4 years.

A significant disadvantage to a chrome recovery system in this particular application, however, is the relatively high concentration of nitrate in the regenerant solution. The recovered acid would thus have a high nitric acid content and could be reused only in the deox baths. Returning acids reclaimed from a mixed waste to a process tank also entails production risks that are high, relative to the projected savings. Thus, a decision was made to defer installation of an acid recovery system until actual waste production could be better quantified and potential effects on production evaluated.

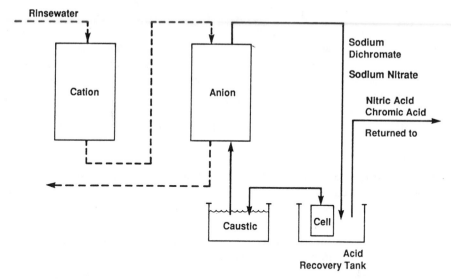

Figure 8.10. Chromic acid recovery system schematic.

Table 8.20. Estimated Savings Due to Installation of Acid Recovery System

Item	Cost (1988 $)
1. Regenerant chemicals (8,000 lb H_2SO_4)	2,400
2. Power for chrome recovery (2 kW)	(1,200)
3. Evaporation power (no reduction)	0
4. Disposal cost (70% reduction)	7,100
5. Bath chemicals savings	2,000
6. Total cost savings per year	10,300
7. Installed cost of acid recovery system	40,000
8. Payback period (years)	3.9

TOXICITY REDUCTION

Disposal costs are considerably higher for disposal in a hazardous waste disposal facility versus disposal in an industrial solid waste disposal facility. Thus, there is a strong economic incentive to treat hazardous waste to render it nonhazardous.

A waste can be rendered nonhazardous by changing the characteristics of the waste that make it hazardous (i.e., corrosivity due to high or low pH can be eliminated by neutralization). An effective method is to recover and recycle the constituent that would render the waste hazardous (see recovery of a useful product in an earlier section of this chapter). Tying up a hazardous constituent in a nonleachable solid matrix (stabilization or solidification) can permit the mixture to pass the appropriate leaching test and thus be classified as nonhazardous. Another treatment method is destruction of the hazardous constituent in a waste mixture (usually associated with destruction of an organic contaminant in a chemical or thermal reaction).

The following case study illustrates how combining waste streams that are classified as hazardous because of corrosivity can result in their conversion to wastes that are classified as nonhazardous.

Case Study 8.5: Mixing Foundry Wastes to Produce a Nonhazardous Waste

Background and Objectives.   A casting company operates three foundry facilities in a metropolitan area in the Pacific Northwest. Two of the foundries (Plants A and C) are located on adjacent sites in the city, and the third (Plant B) is located in a nearby town. Each of these plants uses corrosive materials that require special disposal when spent.

Plants A and B use a strong (12 N) potassium hydroxide (KOH) solution for removing shell material from castings. As the KOH is used, solids build up in the form of a sandy KOH sludge. Spent KOH solutions are pumped out in bulk and sold as a neutralizing agent or disposed of as a caustic liquid waste in a chemical waste disposal facility. Sludge remaining after the liquids are removed is drummed for disposal at the chemical waste disposal facility.

Plant C uses a solution of hydrofluoric and nitric acids to chemically mill (Chem Mill) cast titanium parts. Spent acid is neutralized with purchased sodium hydroxide (NaOH), producing a sludge. Neutralized liquids are decanted to a sanitary sewer, and the sludge is disposed of in an industrial landfill. Plant C also uses a 9 N NaOH solution to remove shell material from castings. The shell material contains a radioactive constituent that causes the NaOH sludge to be classified as a low-level radioactive waste, thus requiring disposal at the Hanford radioactive waste disposal facility. The NaOH liquid is not considered to be a radioactive waste and is sold to local industry as a neutralizing agent.

The casting company recognized the desirability of using its spent alkaline material to neutralize spent Chem Mill acid waste instead of purchasing virgin sodium hydroxide and then paying to dispose of its spent alkaline materials. However, tests showed that mixing the wastes resulted in precipitation of solids. Since the facilities did not have the equipment to deal with the solids, they had not implemented such a program.

The company retained CH2M HILL to develop a conceptual design of a facility capable of neutralizing its spent acids with its own spent caustic materials and handling the resulting solids. A key objective of this study was to determine if the facility would require an RCRA permit as a treatment, storage, and disposal facility, and whether the resulting neutralized solids could be disposed of in an industrial landfill, instead of an RCRA facility.

Laboratory Testing.   Samples of the spent chemicals (and mixtures) were characterized (Table 8.21). Since all of the materials fell outside the pH range of 2 to 12.5, they were considered to be RCRA hazardous wastes. In addition, three of the spent samples exceeded the extraction procedure (EP) toxicity limits for certain metals.

Table 8.21. Foundry Spent Chemical Characterics

Characteristic	Plant A KOH Sludge	Plant A KOH Liquid[a]	Plant B KOH Sludge	Plant B KOH Liquid	NaOH[b] Sludge	Plant C NaOH Liquid	Plant C Chem Mill Solution	Combined KOH[c] Sludge and Chem Mill Solution	Combined KOH[d] Liquid and Chem Mill Solution	RCRA Limits[e]
Total Solids (%)	86	—	85	—	65	—	—	—	23	—
Wet Density (g/mL)	2.19	—	2.16	1.51	2.77	—	1.11	1.53	1.19	—
pH	~14	—	~14	~14	~14	~14	~1	~11	~7	2–12.5
Normality (eq/L)	6.5	—	7.7	11.8	4.2	8.9	3.4	—	—	—
EP Metal Concentration[f] (mg/L)										
Arsenic (As)	< 0.005	—	0.310	1.40	—	0.38	4.60	0.28	0.37	5
Barium (Ba)	< 1.50	—	0.11	6.77	—	< 9.4	2.0	< 1.0	< 1.3	100
Cadmium (Cd)	0.0008	—	0.0005	0.01	—	1.69[g]	0.150	0.022	0.029	1
Chromium (Cr)	< 0.005	—	< 0.005	6.77[g]	—	2.81	21.0[g]	< 0.10	< 0.13	5
Lead (Pb)	< 0.125	—	< 0.125	6.18[g]	—	< 0.09	0.32	< 0.10	< 0.13	5
Mercury (Hg)	< 0.00010	—	0.00028	0.0022	—	0.0015	0.0006	0.0004	0.0005	0.2
Selenium (Se)	0.450	—	0.540	15.1[g]	—	< 0.09	< 0.1	0.24	0.32	1
Silver (Ag)	0.029	—	0.100	0.10	—	0.094	< 0.02	< 0.02	< 0.03	5

[a]Plant A KOH liquid was not analyzed. It was assumed to be similar to Plant B KOH liquid.

[b]Plant C NaOH sludge was not analyzed for solids or metals because of its radioactive character. Solids determination would have concentrated the radioactive material, and the EP metals determination would not have contributed any significant information.

[c]The Plant A KOH sludge and Plant B KOH sludge were combined at a 5:1 ratio and then neutralized with Chem Mill solution to pH 11.

[d]The metals concentrations are 1.32 times those for the KOH sludge and Chem Mill solution combined. The only difference in these two materials is the sand in the sludge, which was assumed not to contribute any metals.

[e]Limits exist under RCRA for determining characteristics of corrosivity and extraction procedure (EP) toxicity.

[f]Metal concentrations shown for Plant A KOH sludge and Plant B KOH sludge are for the filtrate after neutralization.

[g]Metal concentration is higher than EP toxicity standard.

Mixing KOH wastes with the Chem Mill acids produced a solid with a pH range of 7 to 11, which passed the EP toxicity test for heavy metals and therefore was suitable for disposal in a conventional landfill.

Regulatory Issues.   RCRA provides that materials used as a substitute for a pure commercial product are not considered a solid waste and are therefore exempt from RCRA regulation. The preamble to the regulation that allows this exemption discusses preventing shams, in particular, the use of materials that are not nearly as effective as the commercial product or are so highly contaminated with toxic materials that their use is primarily for co-mingling and dilution. In this company's application, the KOH wastes are as effective as the commercial products they are replacing and already are being sold as a commercial product; therefore, they should be eligible for exemption.

Process Description.   Table 8.22 shows the quantities of chemicals produced at the three facilities and relates the caustic materials to the quantities of acidic waste they can neutralize. As can be seen, an excess of caustic liquids will be available for sale.

Figure 8.11 is a schematic for the neutralization/solidification facility. The system is designed to be batch operated because of the small quantities produced. Batch neutralization systems are also easier to control than continuous ones. NaOH sludges at Plant C will be solidified and will continue to be shipped to Hanford for disposal as low-level radioactive waste.

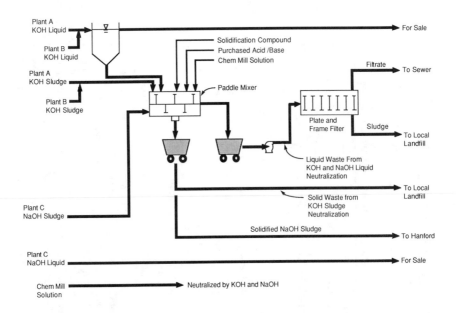

Figure 8.11. Schematic for foundry waste neutralization facility.

Table 8.22.   Foundry Spent Chemical Quantities and Balance

Quantity	Plant A KOH		Plant B KOH		Plant C		
	Sludge	Liquid	Sludge	Liquid	NaOH Sludge	NaOH Liquid	Chem Mill Solution
Batch size (gal)	14	6,000	53	1,670	220	4,800	9,000
Frequency (batches/yr)	780	10.3	780	26	4	9	29.3
Annual amount (gal)	10,800	62,000	41,300	43,400	900	43,200	264,000
Chem mill neutralization capacity (gal)	10,800	186,000	41,300	130,000	900	130,000	—

Acid/Base Balance
Excess caustic liquid: 78,000 gal/yr (available for sale)

Figure 8.12 is a layout of the neutralization facility. Figure 8.13 is a schematic of the neutralization and solids handling equipment. Table 8.23 lists the major systems components and the basis for their sizing. Table 8.24 shows the production rates for waste products from the facility.

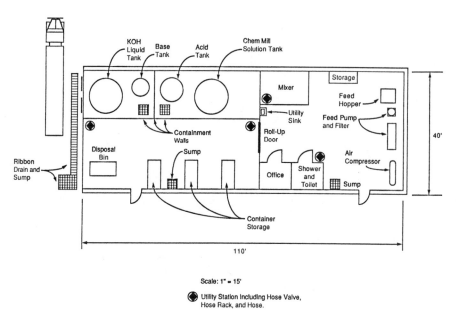

Figure 8.12. Foundry waste neutralization facility layout.

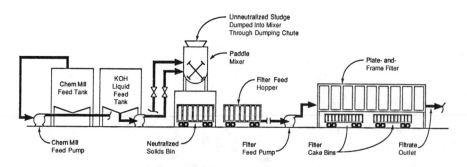

Figure 8.13. Flow schematic for foundry waste neutralization and solids-handling equipment.

Table 8.23.   Foundry Waste Neutralization System—Basis of Design

Treatment System Components	Size	Sizing Basis
KOH liquid tank volume	10,000 gal	Tank sized to hold KOH liquid produced in 4 wk (maximum batch period) plus 700 gal extra volume
Chem Mill solution tank volume	12,000 gal	Tank sized to hold Chem Mill solution produced in 2-wk batch period plus 3,000 gal extra volume
Supplementary acid tank volume	5,000 gal	Tank sized to hold enough 12N supplementary acid to neutralize 2 wk production of KOH waste liquid and sluge
Supplementary base tank volume	6,000 gal	Tank sized to hold one batch of NaOH liquid plus 1,200 gal extra volume
Mixer volume	50 ft³	Mixer sized to hold two 55-gal containers of Plant B sludge, one 55-gal container of Plant A sludge, and the Plant C Chem Mill solution required to neutralize the sludges, with approximately 6 ft³ of extra volume
Neutralized solid bin volume	50 ft³	Bin sized to hold one mixer volume
Filter feed hopper	50 ft³	Hopper sized to hold one mixer volume
Plate and frame filter filtering volume	4 ft³	Filter sized to process the daily production of neutralized KOH liquid in 6 hr
filter area	90 ft²	
Building area	4,400 ft²	
Forklift capacity	5,000 lb	

Table 8.24.   Foundry Waste Neutralization System—Waste Products

Waste Products	Rate	Production Rate Basis
Filtrate	950 gal/day	
Maximum fluoride in filtrate	145 lb/day	The amount of fluoride shown assumes that only the Chem Mill solution is used to neutralize the liquid KOH to pH 7. The resulting fluoride concentration in the plant effluent will be approximately 25 mg/L at peak flow. Filtrate concentration will be approximately 18,400 mg/L fluoride.
Filter Cake		
Solids content	60%	
Volume	700 yd³/yr	
Weight	900 tons/yr	
Neutralized KOH sludge		
Volume	500 yd³/yr	
Weight	630 tons/yr	
Solidified NaOH sludge		
Volume	16 barrels/yr	Amounts shown assume the sludges are mixed with water, then solidified with bentonite clay.
Weight	9 tons/yr	

Spent liquid wastes are transported to the facility in tanker trucks. Chemical sludges are collected in dumpable containers transported on flatbed trucks. Liquids are pumped from the tanker trucks into storage tanks, and sludges are stored in their shipping containers. Radioactive sludges are segregated.

KOH liquids are pumped from the storage tank to the mixer and mixed with Chem Mill acid waste until the waste is neutralized. The resulting thick slurry is collected in a hopper and dewatered with a plate-and-frame filter press. The filtrate is disposed of in a sanitary sewer, and the filter cake is dumped into a bin for disposal in a local landfill.

KOH sludges are brought to the mixer with a forklift equipped with rotating grippers for dumping the containers. Chem Mill acid waste is added to the sludge until the mixture is neutralized. When the mixture does not form a thick solid, a solidification agent, such as bentonite, is added to produce the desired consistency. The neutralized sludge is then collected in the hopper and disposed of with the filter cake.

NaOH liquid that is not sold as a commercial product is stored in the supplementary base tank and used as an alternative neutralization agent replacing KOH. NaOH sludges are processed in the equipment, but separately from all of the other materials, since they are radioactive. After being used for NaOH sludge, the equipment is cleaned to keep the radioactive sludge from contaminating other processed materials.

Capital and Operating Costs.   Table 8.25 shows a budget-level estimate of approximately $700,000 (in 1986 dollars) for the construction costs, including engineering, of the neutralization facility. Table 8.26 shows annual operating costs for the system and compares those costs with the annual costs of treating and disposing of spent chemicals. As shown in the table, the neutralization system was estimated to cost $179,000 per year to operate, for an annual savings of $137,000, which would result in a 5-year payback of the initial investment.

Table 8.25.   Cost Estimates for Foundry Waste Neutralization Facility

Basic Equipment	Size	Cost
Paddle mixer[a]	50 ft^3	$ 55,000
Plate and frame filter[b]	4.1 ft^3	15,000
Tanks[c]		
Chem mill solution	12,000 gal	24,000
KOH liquid	10,000 gal	9,000
Supplementary acid	5,000 gal	10,000
Supplementary base	6,000 gal	5,000
Forklift attachment[d]		6,000
Process pumps[e]		8,000
Sump pumps[f]		35,000
Miscellaneous equipment[g]		36,000
Equipment installation		33,000
TOTAL BASIC EQUIPMENT		$236,000
Site Work and Building		
Site preparation		$ 3,000
Yard piping[i]		11,000
Roads and walks		3,000
Site improvements		3,000
Concrete		21,000
Metal building (4,400 ft^2)[j]		73,000
Miscellaneous building appurtenances[k]		25,000
Miscellaneous metals		5,000
TOTAL SITEWORK AND BUILDING		$144,000
Process piping		$ 18,000
Electrical (approx. 10%)		40,000
Instrumentation (approx. 10%)		40,000
Mobilization/demobilization (approx. 5%)		20,000
Subtotal		$498,000
Contingency (20%)		100,000
TOTAL CONSTRUCTION COST		$598,000

(cont.)

Table 8.25.   Continued

Basic Equipment	Size	Cost
Estimated Engineering Services Cost		
Geotechnical investigation		$ 7,500
Engineering design		73,600
Services during construction		20,000
TOTAL ENGINEERING SERVICES COST		$101,100
TOTAL FACILITY CAPITAL COST		$699,100

[a]The paddle mixer is constructed of stainless steel without a cooling jacket. It will be equipped with a 30-hp motor.

[b]The plate-and-frame filter press is manual-type hydraulically operated and equipped with an air manifold, cake carts, and feed pump.

[c]The Chem Mill solution and acid tanks will be constructed of FRP, and the KOH and base tanks will be constructed of steel.

[d]A forklift equipped with rotating forks will be purchased to move hoppers, bins, and barrels of material.

[e]Air diaphragm pumps are used for corrosion resistance and equipment uniformity. An air compressor will be required for operation of these pumps.

[f] Sump pumps will be installed in the sumps with special pH control loops and level switches.

[g]Miscellaneous equipment includes the hoppers, bins, air compressor, and other minor items.

[h]Equipment installation is estimated at 15% of the equipment cost.

[i] Water and sanitary sewer piping will be brought to the structure from a street approximately 250 ft from the building.

[j] The metal building will enclose only the processing area. The storage area will be provided.

[k]HVAC equipment will include an exhaust fan for the mixer, but not a fume scrubbing system. Only the office and restroom areas will be heated or air conditioned. The building will be ventilated, and freeze protection will be included as required.

Table 8.26.   Cost Savings, Proposed Foundry Waste Neutralization Facility

Item	Annual Cost Current System	Annual Cost[a] Proposed System
Labor[b]	—	$ 45,000
Maintenance[c]	—	25,000
Forklift rental	—	15,000
Materials[d]	—	19,000
Hauling	$ 58,000	23,000[e]
Disposal charge	258,000	36,000[e]
Utilities	—	16,000
TOTAL	$316,000	$179,000

[a]Costs based on material quantities provided by client in December 1986.

[b]Assumes 1-1/2 operators to run the system at $30,000 per year per operator.

[c]Assumes 10 percent of original equipment cost per year.

[d]Includes 40,000 gallons of supplementary acid and 40,000 pounds of solidifying compound per year.

[e]Assumes disposal of filter cake and neutralized KOH sludge at industrial landfill and NaOH sludge at Hanford.

VOLUME REDUCTION

Volume reduction can significantly reduce the handling and disposal costs for residues remaining after treatment, since transportation and disposal costs are usually based on weight or volume. In addition, the presence of free water in an industrial wastewater treatment plant sludge may cause the disposal company to require that kiln dust be added to the waste to make it acceptable to landfill, thus increasing the volume and disposal costs.

Volume reduction at a treatment plant can be accomplished in several ways. Four alternatives include modifications to treatment to reduce the generation of solid residues, sludge thickening, sludge dewatering, and sludge drying.

Reducing Generation of Solid Residues

Waste treatment modifications can result in the reduction of solid residues requiring disposal.

In conventional treatment of a mixed metal waste containing hexavalent chromium, the pH of the waste stream is reduced to below 3 with a mineral acid (usually sulfuric); a reducing agent (sulfur dioxide or sodium metabisulfite) is added to convert the hexavalent chromium to the reduced trivalent state. After the reaction is complete, lime is added to raise the pH of the combined wastewater to approximately 9.5 for precipitation of heavy metals as hydroxides. This treatment scheme, however, can produce a large quantity of sludge, in excess of the metals targeted for removal, because calcium carbonate and calcium sulfate precipitate, particularly when treating a hard alkaline wastewater.

In research sponsored by the Air Force,[2] it was found that chrome reduction could be accomplished at neutral to slightly alkaline pH using a combination of ferrous sulfate and sodium sulfide as agents and sodium hydroxide for heavy metals precipitation. The study found that the resulting iron hydroxide precipitate was effective at removing other heavy metals, such as cadmium and nickel, at neutral to slightly alkaline pH. The usefulness of these modifications includes eliminating an acidic chrome reduction step, eliminating the addition of acid and reducing the need for lime, and precipitating metals without also producing calcium sulfate or carbonate solids.

Example. This process has been adopted at the Tinker Air Force Base industrial wastewater treatment plant to remove chrome, copper, nickel, cadmium, and other metals from a 1-mgd combined industrial waste stream. Operating at a pH of 7.5 to 8.5, the treatment system achieves the same chrome reductions as the old process operated at a pH of 2.5 to 3, while reducing the sludge volumes by two-thirds. The Air Force estimated that this process modification would save $1,000 per day in chemical and sludge disposal costs.[3]

In a similar treatment process, ferrous iron is generated electrochemically using sacrificial iron electrodes. Equipment has been installed at a Navy ordnance plant in Pomona, California, to remove copper and traces of chromium and nickel.[4] The pH of the reactor is held to a range between 6 and 9 using caustic soda (sodium hydroxide). The precipitated metals are settled out in a clarifier and dewatered in a filter press. The process was reported to produce 75% less sludge when compared with acidic chrome reduction and lime precipitation.

Sludge Thickening

Combined industrial wastewater treatment facilities typically employ hydroxide precipitation for removal of toxic metals. The metal hydroxide solids are usually removed by clarification. Metal hydroxide sludges withdrawn from a clarifier typically have solids contents ranging from 1% to 2%. A pound of copper precipitated with hydroxide produces 1.54 lb of solid copper hydroxide. In a 1% sludge, this pound of copper produces 154 lb of sludge. Disposal of such a large volume would be expensive, even if permitted (disposal of liquid wastes in hazardous waste landfills is banned). Therefore, most waste treatment plants dewater their sludges prior to disposal.

Prior to dewatering, use of a thickener can prove useful. First, many small waste treatment plants are built using parallel plate clarifiers to reduce the size of the plant and the cost of treatment. Small packaged clarifiers usually provide limited sludge storage volume and lack a mechanism to scrape sludge to the outlet. As a result, problems are often created because of their inability to draw off sludge completely or to generate a reasonably thick sludge. Adding a thickener at the bottom of a packaged clarifier provides additional sludge storage volume in addition to increasing the sludge solids to as much as 5% to 6% (reducing the 154 lb of copper sludge to 30 lb).

Increasing solids by thickening assists subsequent sludge dewatering in two ways. First it reduces the time required for dewatering, and second, it usually results in a drier sludge cake after dewatering.

Sludge Dewatering

In the past, sludge drying has been accomplished in open sand drying beds, especially in areas of the country with warm climates, low rainfall, and cheap land. However, since sludges from industrial wastewater treatment plants are frequently hazardous, sludge drying beds have all of the design (and potentially regulatory) requirements of a hazardous waste landfill, with collection and treatment of leachate required. Therefore, mechanical dewatering is most frequently used because of regulatory and cost advantages.

Three types of mechanical dewatering devices are typically employed for industrial wastewater treatment plant sludges: vacuum filters, belt presses, and plate-and-frame filter presses.

Vacuum filters are not often used, since they produce the least dry cake of the three mechanical methods. Vacuum filters frequently employ a precoating process, using diatomaceous earth, to improve dewatering. A portion of the precoat is scraped off, adding to the volume of solids to be disposed of. Also, vacuum filters generally require higher energy use than the other mechanical dewatering processes. In some applications (dewatering aluminum hydroxide sludges), vacuum filters are preferred, since they are gentler to gelatinous sludges and can produce a dry cake without blinding the filter cloth, as occurs with the other filters.

Capital costs for the smallest belt press are greater than for the smallest plate-and-frame press. However, belt presses can be economical for large waste treatment plants. Operating costs are lower because of the continuous operation as opposed to the batch operation of a plate-and-frame press. Belt presses generally produce a cake that ranges from 20% to 30% solids.

Plate-and-frame filter presses are generally used for facilities that produce a small volume of waste, since these devices are simple and are available in a broad range of sizes. Sludge from the thickener or clarifier is pumped, typically using an air diaphragm pump, to the chambers of a filter press. The solids are retained by the filter media (polypropylene or other synthetic cloth), and the liquid flows through the media and is returned to the wastewater treatment plant for retreatment. After the pressure required to pump sludge to the press reaches a maximum value, the hydraulic press that holds the plates together is released, and a filter cake is discharged. Plate-and-frame filter presses generally produce the driest sludge cakes (30% to 40% solids), an advantage when disposal is based on weight or volume.

Filter press operation generally requires little operator attention, except at the beginning and end of a cycle. Automatic plate shifters greatly reduce the manual labor required to remove the filter cake. Adequately dewatered sludge will literally fall out of the press when it opens.

The major maintenance cost of a mechanical sludge filter is replacing filter cloths, especially when handling abrasive wastes having an extreme pH. However, metal hydroxide sludges are generally of moderately alkaline pH and nonabrasive.

A list of suppliers of sludge dewatering equipment is provided in Table 8.27.

Table 8.27. Suppliers of Sludge Dewatering Equipment

Company	Address	Phone
Atlantes Chemical	303 Silver Spring Rd Conroe, TX 77303	(409) 856-4515
Avery Filter Co.	99 Kinderkamack Rd Westwood, NJ 07675	(201) 666-9684
BC Hoesch Industries	Richard Mine Rd Wharton, NJ 07885	(201) 361-4700
Industrial Filter	5904 Ogden Ave Cicero, IL 60650	(312) 656-7800
JWI, Inc.	2155 112th Ave Holland, MI 49423	(616) 772-9011
Met-Pro	163 Cassell Rd Harleysville, PA 19438	(215) 723-6751
Netzsch	119 Pickering Way Exton, PA 19341	(215) 363-8010

(cont.)

Table 8.27.   Continued

Company	Address	Phone
OMI International	21441 Hoover Rd Warren, MI 48089	(313) 497-9100
Shriver	EIMCO, P.O. Box 300 Salt Lake City, UT 84110	(801) 526-2000

Sludge Drying

Mechanical dewatering can generally increase the solids content of a sludge from 30% to 40%. Using the 1 lb of copper as an example, an unthickened 1% sludge weighing 154 lb is reduced to 5 lb of filter cake (30% solids). This filter cake still contains 3.5 lb of water. In addition, even though this sludge appears to be dry on the surface, free water typically will escape during shipment. Landfill operators may then either reject the sludge (because of a ban on the landfill disposal of free liquids) or else require that cement kiln dust be added to react with or soak up the excess moisture.

Example. One of CH2M HILL's clients generated a metal hydroxide sludge cake, which was dewatered using a plate-and-frame filter press, producing a sludge cake of 40% solids. The sludge was transferred to 55-gal drums and shipped to a hazardous waste landfill for disposal. At the landfill the drums were opened. On finding free water at the surface, the landfill operators added cement kiln dust to the waste and charged an additional $150 per barrel for the service. The client is presently adding kiln dust prior to shipping, thus doubling the volume being disposed of.

Further drying of the filter cake to 80% to 90% solids content is feasible. The process would reduce the hypothetical sludge weight from 5 lb to 2 lb (80% solids) and eliminate the need for addition of kiln dust to prevent the generation of free water during shipment.

A list of suppliers of sludge dryers is provided as Table 8.28.

INCINERATION TO REDUCE VOLUME, TOXICITY, AND MOBILITY

Introduction

Solid wastes are generated by virtually every commercial enterprise. Grocery stores generate large volumes of rubbish and garbage from packaging, spoiled produce, and out-of-date perishables. The pharmaceutical industry generates waste solvents when purifying ethical drugs and discarding manufactured products or raw materials that do not meet specifications. Paper mills generate waste from wood pulping, trimmings from the paper-making machines, and from roll ends. Dry cleaners generate

Table 8.28. Suppliers of Sludge Dryers

Company	Address	Phone
Atlantes Chemical	303 Silver Spring Rd Conroe, TX 77303	(409) 856-4515
Baker Brothers	44 Campanelli Parkway Stoughton, MA 02072	(617) 344-1700
Fenton Co.	1608 North Beckley Lancaster, TX 75134	(214) 228-3556
BC Hoesch Industries	Richard Mine Rd Wharton, NJ 07885	(201) 361-4700
JWI, Inc.	2155 112th Ave Holland, MI 49423	(616) 772-9011
Sonodyne Ind.	11135 Southwest Capital Highway Portland, OR 97219	(503) 245-7259
Techmatic, Inc.	133 Lyle Lane Nashville, TN 37210	(615) 256-1416
Water Management	2300 Highway 70 East Hot Springs, AR 71901	(501) 623-2221

waste oils, dirty filters used to recycle solvents, and spent solvents. In addition, hospitals generate rubbish, waste laboratory samples, operating room wastes, and discarded medicines, and publishers generate wastes which include printing inks, scrap paper, spoiled copy, and discarded publications. The list is virtually endless.

There are techniques for recycling portions of these wastes. Recyclers must create a product that meets minimum standards for commercial acceptability. For instance, to be saleable, recycled lubricating oil must meet standards for lubricating quality and temperature stability set by the Society of Automotive Engineers. Recycled solvents must meet commercial criteria for industrial solvents. Recycled paper must be suitable for cardboard manufacture, writing paper, or cellulosic insulations. If these commercial constraints are not met, the recycled product itself will be discarded as waste.

Currently there are three methods to dispose of wastes: land disposal, which has been and continues to be the predominant method; waterborne disposal with eventual drainage out to sea; and evaporation and dispersal to the atmosphere.

Treatment prior to disposal is desirable to reduce toxicity, mobility, or volume of the waste. Baling and compressing reduce volume, but do not affect toxicity or decrease mobility. Fixation and solidification of solid wastes reduce the mobility, but only with an attendant increase in volume. In fact, the only method that simultaneously reduces volume, toxicity, and mobility is burning in incinerators, boilers, power plants, or supplementing fuel used in cement kilns or light aggregate kilns.

When a waste is burned (yielding ash or exhaust gas), its chemical makeup is altered, and its toxicity is generally reduced. Volume is reduced by evaporation of water, and simple organics in the wastes are oxidized to water and carbon dioxide. Mobility is reduced since only inorganic metals and silicates remain in the ash, which is a solid with minimum volume. The other product of burning is exhaust gas, which contains the water vapor, carbon dioxide, and excess air required to support the combustion process. Exhaust gas still containing entrained particulates, acid gases, and any organics that did not burn will require extensive further treatment to remove these particulates and acid gases and to complete the combustion of the organics.

Process Description

The following description focuses on incineration as a waste management technique. The principles, however, also apply to boilers, industrial furnaces, or light aggregate kilns.

Feed Storage and Metering

The feed to an incinerator must be collected and controlled to remain within the allowable limits of the combustion device. For instance, water represents a large heat sink. Too much water will cool the incinerator too much, and the waste will not be completely burned. The amount of water fed to the incinerator must therefore be regulated. Chlorinated organics fed to an incinerator must also be controlled since chlorine will form hydrochloric acid (HCl) if there is sufficient hydrogen in the gas phase. This formation is controlled by the equilibrium reaction:

$$2HCl + 1/2 \, O_2 = Cl_2 + H_2O$$

HCl can cause corrosion problems, but it can readily be removed from the gas stream either by water absorption or by reaction with an alkaline scrubbing solution. Other important feed characteristics to be considered are:

- ash content
- heating value or heat of combustion
- sulfur content
- phosphorus content
- nitrogen content
- physical state (solid, liquid, or gas)
- melting point
- boiling point
- particle size distribution for solids
- packaging (drums, pails, bulk)
- metals content

An incinerator can be designed to handle wastes with a wide range of physical properties. Once designed and constructed, however, the feeds to an incinerator must be kept within a particular operating range in order for the incinerator to perform adequately.

After the requisite feed control is established, the feed materials should be collected and stored, at least temporarily. Liquids can be collected and stored in tanks. These tanks can also provide surge capacity to allow for some mixing and blending to meet the feed control criteria. Liquid wastes are then pumped to the incinerator at a controlled rate to be atomized and burned. Flow control valves are typically used to control the feed rate.

Solids can be collected in drums, in a pit, or on a tipping floor. Solids must be inspected either on the receiving floor or on the delivery vehicle to ensure that they can be safely and effectively burned in the incinerator. Drums of solids can be loaded onto conveyors and periodically charged into the incinerator. The bulk solids in pits can be picked up with cranes and loaded into hoppers to be charged at a controlled rate. Solids on tipping floors can be picked up with a front-end loader and loaded into hoppers to be charged to the incinerator. A metering conveyor with a variable speed drive can be used to control the amount of material fed to the incinerator.

Gases can be fed to the incinerator directly from the transport vessel or through a pipeline. Gas flow rate can be controlled by a flow control valve.

Combustion Control

The various feeds to the incinerator generally react according to the following formulas:

Combustibles + Air + Water =
Ash + Flue Gases + Water Vapor + Excess Air
and
Heat Input + Heat Generated by Combustion = Heat Out

The operator of the incinerator can control few variables around the combustor. These typically are:

- feed rate of wastes

- excess air

- temperature inside the incinerator

These variables are interactive. Increasing the feed rate of wastes at a given total air flow rate will result in a higher temperature. Increasing the excess air at a constant waste feed rate will reduce temperature. The effects of these variables can be separated by addition of an auxiliary fuel.

By supplementing the fuel, the feed rate and air rate are set and the temperature increased above that which can be achieved by combustion of the waste alone. The auxiliary fuel is then added to create additional heat and raise the temperature of the incinerator to the desired set point. As feed rate of waste, heating value of the waste, and air rates vary about the nominal condition, the flow of auxiliary fuel is varied to maintain constant temperature.

Once the balance of feed rates, air rates, and temperature is achieved, an incinerator operates at steady state, burning away the combustibles in the wastes, reducing the volume of the ash to a minimum, and destroying the organics contained in the waste.

Flue Gas Treatment

The gases created in the incinerator will require treatment to remove particulates, acid gases, and trace organics that may remain from the incineration process. Trace organics are further oxidized either by providing sufficient reaction time at an adequate temperature in the primary combustor or by passing the flue gas from the primary combustion chamber through an afterburner to complete the combustion process.

After combustion, the gas stream is quenched to reduce its temperature, and particles are removed. Several technologies are used in incinerator installations, including:

- venturi scrubbers

- inertial impact scrubbers

- baghouse filters

- electostatic precipitators

- plate scrubbing towers

Acid gases can be removed either by absorption in water to generate an aqueous acid stream or by reaction with an alkaline scrubbing solution to generate a salt stream. Acid gas absorption can be accomplished in:

- venturi scrubbers

- packed tower or tray scrubbers

- multistage absorption towers

All of these technologies have been applied to incinerators.

Scrubbing systems can be either wet or dry. Wet systems generate a wastewater stream, which must be treated to remove suspended solids and perhaps adjust the pH prior to discharge. Dry scrubbers generate a dry solid waste, which can be disposed of with the ash from the incinerator, although the latter systems require closer control and more operator attention.

Thermal Treatment for Resource or Energy Recovery

Thermal processes can be used for recovering mineral values from wastestreams and for recovering energy in the form of steam or electricity.

Chlorinated organic compounds, when burned, will generate HC1 gas and a small amount of chlorine gas. The HC1 can be absorbed in a multistage absorption tower to manufacture hydrochloric acid at strengths varying from 6% to 24% HC1. These acid streams have been used for pickling acid in steel manufacture.

Sulfur-bearing wastes containing more than about 20% sulfur can be burned. The sulfur dioxide can then be catalytically oxidized to sulfur trioxide and absorbed to manufacture sulfuric acid. An alternative process can be used to reduce the sulfur dioxide to make elemental sulfur.

The previous examples are of well established industrial processes that have been modified to recover commercial products from wastes. The use of these methods has been limited, thus far, to situations where the wastestreams are generated in a chemical production plant and the furnace has been added to the production process for waste treatment or air pollution control.

Recently, work has been done on the feasibility of treating ash produced by solids incinerators to recover iron from the ash. These methods have not yet proved fruitful because of the low price of iron. However, reclaiming noble metals from the manufacture of printed circuit boards and electronic parts is being actively considered as an alternative to disposal. As with all recycling/reclamation projects, the attractiveness depends on the value of the reclaimed product and the cost of reclamation.

Recovering energy in the form of steam and/or electricity from a thermal treatment process has been much more widespread. The exhaust gases from the furnace are cooled in a boiler, and the heat is used to boil water under pressure. The typical limit for steam generation in these applications is less than 400 psig, and 150-to-200-psig steam with nominal superheat is common practice. The steam can be exported to nearby consumers or used to drive turbine-generators to generate electricity. Additionally, the steam can be used to chill water through absorption refrigeration units to provide cooling for nearby consumers.

REFERENCES

1. Schick, H. W., and R. M. Rosain, "Vanadium Removal from Oil-Fired Power Plant Wastewaters," presented at the 46th Annual Water Conference, Pittsburgh, PA, November 4-7, 1985.
2. Higgins, T. E., and S. Termaath, "Treatment of Plating Wastewaters by Ferrous Reduction, Sulfide Precipitation, Coagulation and Upflow Filtration," in *Proceedings of the 36th*

Purdue Industrial Waste Conference (Ann Arbor, MI: Ann Arbor Science Publishers, Inc., 1982), pp. 462-471.

3. "A New Waste-Treatment Method is Saving the US Air Force $1,000/day," *Chem. Eng.* (April 28, 1986), pp. 10-11.

4. Roberts, R. M., J. L. Koff, and L. A. Karr, "Hazardous Waste Minimization Initiation Decision Report," Technical Memorandum TM 71-86-03, Naval Civil Engineering Laboratory, Port Hueneme, CA (January 1988).

Acronyms and Abbreviations

AFB	Air Force Base
AFESC/RDV	Air Force Engineering and Services Center/Environics Division
Al	Aluminum
$AlCl_3$	Aluminum chloride
API	American Petroleum Institute
ASTM	American Society for Testing and Materials
Biox	Biochemical oxidation
BOD	Biochemical oxygen demand
BP	Boiling point
Btu	British thermal unit
C	Centigrade
Cd	Cadmium
ft^3	Cubic feet
ft^3/min	Cubic feet per minute
Chem Mill	Chemical milling [facility]
CN	Cyanide [ion or complex]
COD	Chemical oxygen demand
Cu	Copper
yd^3	Cubic yards
yd^3/yr	Cubic yards per year
DAF	Dissolved air flotation
DC	Disposal costs
Deox	Deoxidizer
DI	Deionized [water]
DOD	Department of Defense
ED	Electrodialysis
Eq/day	Equivalents per day
EP	Extraction procedure
EPA	U.S. Environmental Protection Agency
F	Fahrenheit
F.O.B.	Freight on board
FRP	Fiber reinforced plastic
FY	Fiscal year
gal/hr	Gallons per hour

gal/min	Gallons per minute
GOCO	Government-owned and company-operated
GOGO	Government-owned and government-operated
gal/yr	Gallons per year
HCl	Hydrochloric acid
HMT	High mass transfer
hp	Horsepower
HSWA	Hazardous and Solid Waste Amendments
I&C	Instrumentation and control
IVD	Ion vapor deposition
IX	Ion exchange
KOH	Potassium hydroxide
kW	Kilowatt
kWh	Kilowatt-hour
lb/hr	Pounds per hour
LSA	Low surface area
MC	Methylene chloride
MD	Man-days
MEK	Methyl ethyl ketone
MH	Man-hours
meq/mL	Milliequivalents per milliliter
mg/L	Milligrams per liter
MIBK	Methyl isobutyl ketone
MSDS	Material Safety Data Sheets
MTBF	Mean time between failure
MW	Megawatt
NADEP	Naval Aviation Depot
NaOH	Sodium hydroxide
NARF	Discontinued designation; see NADEP
NAS	Naval Air Station
NASA	National Aeronautics and Space Administration
NCEL	Naval Civil Engineering Laboratory
Ni	Nickel
NMP	N-methyl-2-pyrrolidone
NPDES	National Pollutant Discharge Elimination System
NSY	Naval Shipyard
OC	Operating costs
O&M	Operation and maintenance
OH	Hydroxide ion
OSHA	Occupational Safety and Health Administration
OTA	United States Congressional Office of Technology Assessment
Pb	Lead
PERC	Perchloroethylene
pH	Degree of acidity or alkalinity of a solution
PMB	Plastic media blasting

POTW	Publicly owned treatment works
ppm	Parts per million
PRP	Potential responsible parties
psig	Pounds per square inch gauge
R&D	Research and development
RCRA	Resource Conservation and Recovery Act
RFIX	Reciprocating flow ion exchanger
RO	Reverse osmosis
sft³/min	Standard cubic feet per minute
ft²	square feet
SO₂	sulfur dioxide
Sn	Tin
SPC	Statistical process control
SS	Stainless steel
TAP	Technical assistance program
TCA	Trichloroethane
TCE	Trichloroethylene
TDS	Total dissolved solids
TSD	Treatment/storage/disposal [facility]
TSS	Total suspended solids
USAF	United States Air Force
UF	Ultrafiltration
VOC	Volatile organic compounds
WFE	Wiped film evaporator
Wt	Weight
Zn	Zinc

Index